Springer Theses

Recognizing Outstanding Ph.D. Research

Aims and Scope

The series "Springer Theses" brings together a selection of the very best Ph.D. theses from around the world and across the physical sciences. Nominated and endorsed by two recognized specialists, each published volume has been selected for its scientific excellence and the high impact of its contents for the pertinent field of research. For greater accessibility to non-specialists, the published versions include an extended introduction, as well as a foreword by the student's supervisor explaining the special relevance of the work for the field. As a whole, the series will provide a valuable resource both for newcomers to the research fields described, and for other scientists seeking detailed background information on special questions. Finally, it provides an accredited documentation of the valuable contributions made by today's younger generation of scientists.

Theses may be nominated for publication in this series by heads of department at internationally leading universities or institutes and should fulfill all of the following criteria

- They must be written in good English.
- The topic should fall within the confines of Chemistry, Physics, Earth Sciences, Engineering and related interdisciplinary fields such as Materials, Nanoscience, Chemical Engineering, Complex Systems and Biophysics.
- The work reported in the thesis must represent a significant scientific advance.
- If the thesis includes previously published material, permission to reproduce this must be gained from the respective copyright holder (a maximum 30% of the thesis should be a verbatim reproduction from the author's previous publications).
- They must have been examined and passed during the 12 months prior to nomination.
- Each thesis should include a foreword by the supervisor outlining the significance of its content.
- The theses should have a clearly defined structure including an introduction accessible to new PhD students and scientists not expert in the relevant field.

Indexed by zbMATH.

More information about this series at http://www.springer.com/series/8790

Chao Gong

Crash Safety of High-Voltage Powertrain Based Electric Vehicles

Electric Shock Risk Prevention

Doctoral Thesis accepted by
University of York, York, UK

Springer

Author
Dr. Chao Gong
Department of Electronic Engineering
University of York
York, UK

Supervisor
Prof. Yihua Hu
University of York
Heslington, York, UK

ISSN 2190-5053 ISSN 2190-5061 (electronic)
Springer Theses
ISBN 978-3-030-88978-4 ISBN 978-3-030-88979-1 (eBook)
https://doi.org/10.1007/978-3-030-88979-1

This Springer imprint is published by the registered company Springer Nature Switzerland AG
The registered company address is: Gewerbestrasse 11, 6330 Cham, Switzerland

Declaration

I declare that this thesis is a presentation of original work and I am the sole author. This work has not previously been presented for an award at this, or any other, university. All sources are acknowledged as References.

Main contents in Chap. 2 have been submitted to *IEEE Transactions on Industrial Electronics*. Title is "Safety of Electric Vehicles in Crash Conditions: A Review of Hazards to Occupants, Regulatory Activities and Technical Support".

Main contents in Chap. 3 have been published to *IEEE Transactions on Industrial Informatics*, vol. 15, no. 8, pp. 4709–4718, Aug. 2019. Title is "A DC-Bus Capacitor Discharge Strategy for PMSM Drive System with Large Inertia and Small System Safe Current in EVs".

Main contents in Chap. 4 have been published to *IEEE Transactions on Industrial Electronics*, vol. 68, no. 3, pp. 1905–1915, March 2021. Title is "Hybrid DC-Bus Capacitor Discharge Strategy Using Internal Windings and External Bleeder for Surface-mounted PMSM based EV Powertrains in Emergency".

Main contents in Chap. 5 have been published to *IEEE Transactions on Industrial Electronics*, vol. 67, no. 10, pp. 8131–8142, Oct. 2020. Title is "Reliable Winding-based DC-Bus Capacitor Discharge Technique Over Full-Speed Range for IPMSM Drive".

Main contents in Chap. 6 have been published to *IEEE Transactions on Power Electronics*, vol. 36, no. 3, pp. 3551–3562, March 2021. Title is "Winding-based DC-Bus Capacitor Discharge Technique Selection Principles based on Parametric Analysis for EV-PMSM Drives in Post-Crash Conditions".

December 2020 Chao Gong

Supervisor's Foreword

Electric vehicles are promising as clean energy powered vehicles, due to increased interest in fuel economy and environment in recent years. However, it requires to meet unique safety performance such as crash safety and electric safety.

In terms of a permanent magnet synchronous machine (PMSM) drive system which contains six main components (battery pack, converter, DC-bus capacitor, contactor, inverter, and machine), when an emergency like car crash occurs, the powertrain will operate under a protection mode. The protection mode is endowed with the following characteristics: (1) the contactor is tripped; (2) the clutch in the vehicle disconnects the wheels from the traction motor and the motor just rotates with no load; (3) all the power devices in the inverter are switched off. In this case, the electrical energy stored in the capacitor cannot be dissipated and the bus voltage remains at a high level, posing great threat (electrical shock risks) to the passengers. According to the United Nation Vehicle Regulation ECE R94, the DC-bus capacitor voltage needs to get down to a safe level (60 V) fast (5 s) to avert electrical shock risks after EV crash accident. To achieve this goal, the traditional method using bleeder circuits composed of a switch and power resistor in parallel with the capacitor can be adopted to quickly expend both the electrical energy in the capacitor and the residual mechanical energy in the system. However, the extra circuits sacrifice the weight, volume, and cost of the EV drive systems.

As the supervisor of the author (Chao Gong) of this thesis, I recommended to develop a fast DC-bus capacitor discharge strategy without using complicated bleeder circuits. During Chao's PhD life, he studied how to use the internal components (e.g., machine windings) of the EV drive system to discharge the DC-bus capacitor. In this thesis, he has developed systematic fast winding-based discharge strategies, which are applicable to real applications. By using the proposed techniques in his research, EV companies can use more cost-effective methods to make their products conform to standards of UN Vehicle Regulation (ECE R94). In the future, as the safety of EVs is greatly improved, ordinary people will be more willing to embrace the EV, contributing to decarbonization and environment protection. Thanks to this research, the reliability and safety issues of EVs have inspired several EV and motor drive research communities to pay more attention to this issue than the case at present.

This thesis clearly introduces the winding-based discharge techniques developed in recent years. Regarding the errors and inadequacies in the thesis, I sincerely request readers and peers to criticize and correct them.

Heslington, York, UK
July 2021

Prof. Yihua Hu

Abstract

As an environmentally friendly transport option, electric vehicles (EVs) are endowed with the characteristics of low fossil energy consumption and low pollutant emissions, making them a promising substitute for the conventional gasoline vehicles. Currently, because the high-voltage propulsion systems have advantages of high efficiency, compact structure, and low cost, etc., they have been widely employed in EVs. However, the high-voltage powertrain-based EVs are subjected to more complicated safety problems (e.g., electric shock injury hazards) compared to the conventional internal combustion engine (ICE) cars, especially in the crash scenarios. To investigate and improve crash safety of the high-voltage powertrain-based EVs are important for their promotions.

The purposes of this study are to investigate the crash safety-related issues and develop effective safety improvement techniques for EVs. Particularly, the electric shock risks and the corresponding prevention strategies during crash are the main concerns. In the first place, a literature review is carried out to illustrate the safety-related topics from the perspectives of the possible injury hazards that an occupant might suffer and the efforts that have been made to prevent injuries. The review work indicates that it is valuable and urgent to study and develop winding-based DC-bus capacitor discharge techniques for preventing electric shock risks reliably and cost-efficiently. On this ground, this study, then, focuses on the theories and techniques concerning this topic so as to tackle different challenges encountered in the actual applications. In detail, firstly, as for the systems with extreme parameters, a novel nonzero d-axis, non-zero q-axis (NDNQ) current injection-based discharge strategy is designed to quicken the discharge process. Secondly, a hybrid discharge strategy that combines machine windings and external bleeder circuits is proposed for the systems with highly extreme parameters. Thirdly, considering that the components in the EV powertrains are likely to malfunction during crash, a winding-based discharge strategy with fault tolerance against position signal loss is developed. Finally, to solve the problem that it is difficult to pre-evaluate whether or which winding-based discharge technique is available to an EV at the design stage, this study provides analytical selection principles on the basis of parametric analysis.

List of Publications

[1] **C. Gong**, Y. Hu, G. Chen, H. Wen, Z. Wang and K. Ni, "A DC-Bus Capacitor Discharge Strategy for PMSM Drive System with Large Inertia and Small System Safe Current in EVs," *IEEE Transactions on Industrial Informatics*, vol. 15, no. 8, pp. 4709–4718, Aug. 2019.

[2] **C. Gong**, Y. Hu, H. Wen, G. Chen, W. Li, J. Gao, "Reliable Winding-based DC-Bus Capacitor Discharge Technique Over Full-Speed Range for IPMSM Drive,"*IEEE Transactions on Industrial Electronics*, vol. 67, no. 10, pp. 8131–8142, Oct. 2020.

[3] **C. Gong**, Y. Hu, W. Li, J. Gao, J. Liu, H. Wen, J. Yang, "Hybrid DC-Bus Capacitor Discharge Strategy Using Internal Windings and External Bleeder for Surface-mounted PMSM based EV Powertrains in Emergency," *IEEE Transactions on Industrial Electronics*, doi: 10.1109/TIE.2020.2975479

[4] **Chao Gong**, Yihua Hu, J. Gao, Z. Wu, J. Liu, H. Wen, Z. Wang, "Winding-based DC-Bus Capacitor Discharge Technique Selection Principles based on Parametric Analysis for EV-PMSM Drives in Post-Crash Conditions," *IEEE Transactions on Power Electronics*. doi: 10.1109/TPEL.2020.3017719.

[5] **Chao Gong**, J. Gao, Yihua Hu, F. Zhou, "Safety of Electric Vehicles in Crash Conditions: A Review of Hazards to Occupants, Regulatory Activities and Technical Support,"*IEEE Transactions on Industrial Electronics*. (Under Review)

[6] **C. Gong**, Y. Hu, C. Gan, G. Chen and M. Alkahtani, "Modeling, Analysis, and Attenuation of Uncontrolled Generation for IPMSM-Based Electric Vehicles in Emergency," *IEEE Transactions on Industrial Electronics*, vol. 67, no. 6, pp. 4453–4462, June 2020.

[7] **C. Gong**, Y. Hu, J. Gao, Y. Wang and L. Yan, "An Improved Delay-Suppressed Sliding-Mode Observer for Sensorless Vector-Controlled PMSM," *IEEE Transactions on Industrial Electronics*, vol. 67, no. 7, pp. 5913–5923, July 2020.

[8] **C. Gong**, M. Alkahtani, W. Li, T. Wu, Y. Hu and J. Liu, "FCS Model Predictive Current Control Method for EV PMSMs at Low Control Frequency Considering Flux Mismatch," *2020 IEEE Applied Power Electronics Conference and Exposition (APEC)*, New Orleans, LA, USA, 2020, pp. 265–270.

[9] T. Wu, **C. Gong**, J. Gao and J. Lu, "A Winding-based DC-Bus Capacitor Discharge Strategy for PMSM Drive System in EVs Considering Position Sensor Fault," *IECON 2018 - 44th Annual Conference of the IEEE Industrial Electronics Society, Lisbon, Portugal*, 2019.

Acknowledgements

First and foremost, I would like to express my deepest gratitude to my main supervisor, Dr. Yihua Hu, for granting me the precious opportunity to further myself at the highest level. He has encouraged and helped me to progress academically from a fresh graduate to a professional researcher and a mature individual. I have learned so much from him both academically and personally. Thank him for the invaluable comments and advice on my research as well as my life and career. It is a great honor for me to be his Ph.D. student, and I hope I have fulfilled his high expectations and faith through my achievements. Secondly, I would also like to thank my second supervisor Prof. Yongbing Xu for the generous support and academic guidance he has provided me.

I would also like to take this opportunity and thank my parents. They have always been greatly supportive of me. Their continuous help and understanding have made my life full of love, and I am grateful for everything they have done. I feel very sorrowful that my mother passed away on September 11, 2020. Although she cannot witness my graduation, I believe she must be proud of me. I have to thank my mother again, and I miss her much. Additionally, I would like to express my appreciation to my girlfriend Miss Jinqiu Gao. I am very thankful to her for her patience, tolerance, company and warm love. I have to say without her help, I would not have made it this far. I am very grateful to my cooperators Prof. Gaolin Wang (Harbin Institute of Technology), Prof. Jinglin Liu (Northwestern Polytechnical University), Associate Prof. Yaofei Han (Tongji University), Prof. Huiqing Wen (University of Liverpool), Prof. Zheng Wang (Southeast University), Associate Prof. Liming Yan (Chang'an University), Ms. Wenzhen Li and Dr. Tianhao Wu, who have given me many useful suggestions for my research and helped me complete my experimental verifications. Special thanks are also paid to my brilliant and lovely colleagues and friends, in particular to Dr. Kai Ni, Mr. Zuyu Wu, Dr. Sen Song and Mr. Yixuan Zhang for many fruitful discussions and enjoyable moments.

Last but not least, the support from the Department of Electronic Engineering at the University of York is gratefully acknowledged. Especially, I am grateful to Ms. Camilla Danese who is really elegant, patient and kind.

Acknowledgements

Contents

Abbreviations

AC	Alternating current
ADR	Australian design rules
AIS	Automotive Industry Standards
APF	Abdominal peak force
BMS	Battery management system
BR	Bleeding resistor
BV	Balanced voltage
CFRP	Carbon fiber-reinforced plastic
CRVeM	Carbon fiber reinforcement vinyl ester with microsphere
CSF	Continuous sigmoid function
DC	Direct current
DP	Dual phase
DSP	Digital signal processor
ECE	Economic Commission for Europe
EFM	Energy flow model
EMF	Electromotive force
EV	Electric vehicle
FAME Phase II	Faster Adoption and Manufacturing of Electric Vehicles in India Phase II
FFC	Femur force criterion
FML	Fiber metal laminate
FMVSS	Federal Motor Vehicle Safety Standards
FOA	Fly optimization algorithm
GB, GB/T	Guobiao
GHG	Greenhouse gas
GMT	Glass mat thermoplastic
HF	High frequency
HIC	Head injury criteria
HSF	High switching frequency

HSPO	High-speed position observer
ICE	Internal combustion engine
ID	Instrument board
IEA	International Energy Agency
IP	Ingress protection
KETEE	Kinetic energy that is converted into the electric energy
KMVSS	Korea Motor Vehicle Safety Standards
LDA-CI	Large-*d*-axis current injection
LPF	Low-pass filter
LSF	Lateral shoulder force
MSKJ	Movement of the sliding knee joints
NBM	Neck bending moment
NCF	Neck compression force
NDNQ	Non-zero *d*-axis, non-zero *q*-axis
NDZQ	Non-zero *d*-axis, zero *q*-axis
NIC	Neck injury criteria
NSF	Neck shear forces
NSGA-II	Non-dominated sorting in genetic algorithms-II
NTF	Neck tensile force
PAR	Posterior angular rotation
PI	Proportional integral
PL-FT	Position loss fault-tolerant
PMSM	Permanent magnet synchronous motor
PSPF	Pubic symphysis peak force
PWM	Pulse width modulation
RC	Resistance–capacitance
RDC	Rib deflection criterion
RESS	Rechargeable energy storage system
RHA	Resultant head acceleration
RLSA	Resultant lower spine acceleration
RSWD	Residual steering wheel displacement
SM	Sliding mode
SOSM	Second-order sliding mode
SVPWM	Space vector pulse width modulation
SW	Sine wave
SWL	Steering wheel
SW-LSPO	Sine wave-based low-speed position observer
TCFC	Tibia compression force criterion
ThCC	Thorax compression criterion
TI	Tibia index
TRB	Tailor Rolled Blank
UN	United Nations
UR	Uncontrolled rectifier

US	United States
V·C	Viscous criterion
VBP	Voltage balance phenomenon
VDR	Voltage descending rate
WS	Windshield

Nomenclature

a_{dec}	Deceleration
a_{dec_i}	Deceleration in each segmented period
C	DC-bus capacitor
C'_e	Voltage constant
$D_{1,\ldots,6}$	Six diodes in inverters
d_b, l_b, m_b	Diameter, length and mass of bleeding resistor
e_α, e_β	α, β-axis back EMF
$\hat{e}_\alpha$, $\hat{e}_\beta$	α, β-axis estimated back EMF
F	Viscous coefficient
f_1, f_2, f_3	Energy flow in energy flow model
i_A, i_B,i_c	Phase current of motor
I_{safe}	System safe current
i_d, i_q	d, q-axis current
i_b	Bleeding current
i_{bRMS}	Root-mean-square bleeding current
i_{b_exRMS}	Root-mean-square bleeding current for bleeding resistor design
i_{ca}	Current-carrying capacity of conductor
I_r	Required discharge current
i_{dc}	DC component of the bus current
i_{d_ref}, i_{q_ref}	d, q-axis current reference
i_{dref_des}, i_{qref_des}	d,q-axis current refence for bleeding resistor design
$i_{d\_ref\_i}$, $i_{q\_ref\_i}$	d, q-axis current reference in each segmented period
i_{d_dis}, i_{q_dis}	d, q-axis discharge current
i_α, i_β	α, β-axis current
$i_{\alpha h}$, $i_{\beta h}$	α,β-axis high-frequency current
$\hat{\underline{i}}_\alpha$, $\hat{\underline{i}}_\beta$	α, β-axis estimated current
$\bar{i}_\alpha$, $\bar{i}_\beta$	α, β-axis current errors
J	Rotor inertia

L_s, L_d, L_q	Phase inductance and$d$, q-axis inductance
N_a, N_b	Number of winding turns
p	Number of pole pairs
Q_0	Initial energy in DC-bus capacitor
Q_{rem}	Remaining energy in capacitor for energy flow model
Q_b	Total energy that needs to be dissipated
Q_{b_des}	Total energy for bleeding resistor design
Q_{bex_des}	Energy dissipated by external bleeder for bleeding resistor design
Q_{bin_des}	Energy dissipated by machine windings for bleeding resistor design
Q_{bin}	Energy dissipated by machine windings
Q_{bex}	Energy dissipated by external bleeder
Q_{safe}	Remaining energy in capacitor at safe voltage
Q_{bfri}	Energy dissipate by friction
Q_{tot}	Energy can be dissipated by machine windings and friction
R_b	Bleeding resistor
R_s	Resistance and inductance of phase windings
s	Controllable switch in bleeding circuits
S_α, S_β	Sliding surfaces
t	Time
t_r	Required discharge period
Δt	Segmented period for a piecewise NDNQ discharge method
T_e	Electromagnetic braking torque
T_{e_i}	Electromagnetic braking torque in each segmented period
$T_{1,\ldots,6}$	Six transistors in inverter
U_a, U_b, U_c	Phase voltage of motor
U_{dc}	Real-time DC-bus voltage
U_{dc0}	Initial DC-bus voltage
u_d, u_q	d, q-axis control voltage
U_{safe}	Safe DC voltage
U_{ref}	DC-bus voltage reference for voltage-regulation based discharge
u_α, u_β	α, β-axis voltage
$u_{\alpha h}$, $u_{\beta h}$	α, β-axis high-frequency voltage
U_h	Amplitude of high-frequency voltage
U_{hr}	Compensated amplitude of high-frequency voltage
v_1, v_2	Values in energy flow model
$\omega_{m\_iq\_st}$, $\omega_{m\_iq\_ed}$	Rotor speeds before and after q-axis current is injected into the machine
ω_m	Rotor speed
ω_{rated}	Rated speed
ω_{m0}	Initial speed
ω_{m_th}	Threshold speed at which the back EMF of PMSM is U_{safe}
ω_{ref}	Reference speed

ω_{m_i}	Rotor speed at the end of each segmented period
$\omega_{pos.\ th}$	Switching speed for high-speed observer and low-speed observer
ω_{m_tr}	Speed after five-second discharge
ω_h	Frequency of high-frequency voltage
$\hat{\omega}_m$	Estimated speed
ρ	Perturbation term in sliding mode observer
ρ_r, ρ_m	Resistivity and density of conductor
λ_1, λ_2	Coefficients in sliding mode observer
$\xi_{min,}$ ξ_{max}	Upper and lower limit of the current estimation errors
θ	Rotor position
$\hat{\theta}$	Estimated position and speed
Ψ_f	Permanent magnet flux linkage

List of Figures

List of Tables

List of Accompanying Materials

Chapter 1
Introduction

1.1 Motivation

In comparison with the traditional gasoline internal combustion engine (ICE) vehicles, electric vehicles (EV) are characterized by no fossil energy consumption and low/zero pollutant emissions, so many governments have regarded them as the future direction of the automotive industry [1–4]. Table 1.1 shows the up-to-date supporting policies and targets (EV market share by 2030, excluding two/three wheelers) for EVs from five main countries/regions, which will have significant impacts on the EV market in the near future [5]. It can be seen that, from a perspective of the governmental policies, tightening greenhouse gas (GHG) emission and standardizing the fuel economy are the most popular means to boost growth of EVs. Besides, special programs are employed regionally. For example, India approved the proposal for implementation of the "Faster Adoption and Manufacturing of Electric Vehicles in India Phase II" (FAME Phase II) scheme in February 2019. In terms of the industrial policies, the technological revolutions in batteries, electric drives and chargers, etc., are highly valued. Under the incentive and guidance of these policies, the number of EVs will increase significantly in the near future. Specifically, in 2019, the global sales of electric cars topped at 2.1 million, boosting the worldwide stock of EVs to 7.2 million [6]. As is shown in Fig. 1.1, the New Policies Scenario released by the International Energy Agency (IEA) projects a global stock of pure EVs of 13 million vehicles by 2020 and nearly 130 million vehicles by 2030 [7].

An increasing number of EVs are now flooding into our daily life, and this draws much attention from not only the professionals working in the relevant fields but also the ordinary consumers. At present, one of the most highlighted concerns is the EV safety problem. It is reported that EV accidents have killed hundreds of people over the past 10 years. For example, a taxi driver died from high-voltage electric shock in 2016 in Taiyuan, China [8]; Because battery packs caught fire and exploded, five people died in an EV accident in May 2018 after they drove their Tesla Model S for only one month [9]. Those terrible accidents motivate us to ask the following question '*is it possible to attain a satisfying EV safety level which is*

C. Gong, *Crash Safety of High-Voltage Powertrain Based Electric Vehicles*,
Springer Theses, https://doi.org/10.1007/978-3-030-88979-1_1

Table 1.1 Supporting policies and targets from five countries/regions

Regions	Governmental regulations	Industrial policies	EV market share by 2030 (%)
EU 	(1) Enact tightened GHG emission standards for vehicles (2) Mandate public procurement for EVs (3) Announce ICE bans	• Launch the European Battery Alliance to promote the battery industry	23
USA 	(1) Freeze greenhouse gas emission standards (2) Announce EV mandate in ten states (3) Adhere to standards for fuel economy	• Funds battery and electric drive R&D by Energy's Vehicle Technologies Office	8
China 	(1) Tighten average fuel economy (2) Prohibit investments in new ICE production plants from 2019 (3) Enact standards for fuel economy	• Require the manufacturers to produce a certain share of EVs	26
Japan 	(1) Enact standards for fuel economy	• Reduce 80% of greenhouse gas emissions per vehicle	21
India 	(1) Enact GHG emission standards (2) Implement FAME Phase II	• Establish public chargers in cities and along highways	11

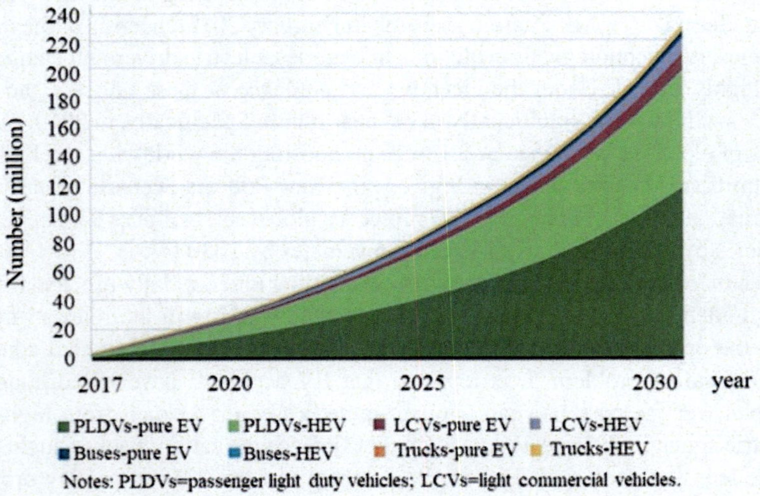

Fig. 1.1 Number of EVs in the future according to EV30@30 scenario [7]

no worse than the traditional ICE cars?'. Safety issues of EVs are closely related to millions of consumers' personal safety and they are important factors restricting the development of EVs. Now, 'without safety, no future for EVs' has become the general industry consensus [10]. Hence, to thoroughly improve the intrinsic safety of EVs is a prerequisite for their promotions.

Crash is the most crucial factor that contributes to road casualties [11, 12], so crash safety is one of the top concerns for EVs at present. The crash safety of EVs differs from that of the conventional vehicles due to the lack of ICEs but the addition of battery-powered propulsion systems. Once an EV crashes or is hit when moving/parking on the road, in addition to the general collision risks that can occur to an ICE car, it is possible for the EV to experience the issues of electrolyte leakage, short circuit and fire, etc., posing more severe risks to the occupants and the responders [13]. Therefore, for EVs, apart from the protection needs that are similar as the general cars, they must also meet extra electrical safety requirements in the crash conditions. For example, with regards to protection against electric shock for the high-voltage powertrain-based EVs, the Economic Commission for Europe (ECE) of the United Nations (UN) Regulation R94 specifies that the voltages in the vehicle components require to drop to the safe level ($\leq$ 60 VDC or 30 VAC) fast (5 s is the best and 60 s is the worst) after crash [14, 38]. Considering the particularity of the high-voltage powertrain-based EVs, it is necessary and urgent to study their crash-related issues and develop effective crash safety improvement techniques.

This research focuses on investigating the crash safety-related issues and developing novel crash safety improvement techniques (especially electric shock prevention techniques) for the high-voltage powertrain-based EVs. The progressive objectives of this study can be summarized as follows:

(i) To **review the safety-related topics of high-voltage powertrain-based EVs in the crash conditions from the perspectives of possible hazards that an occupant is subjected to and efforts made to prevent injuries**. Apart from the electric shock hazards, the physical injuries, corrosion, intoxication and burn injury hazards are surveyed. Further, both the international and national regulations that are enacted to deal with these hazards are presented. Moreover, the up-to-date technologies that can be adopted to reduce injury hazards in the crash conditions are discussed. Through the review work, the necessity of studying the crash safety of EVs is addressed. Besides, a valuable topic about electric shock prevention by developing novel winding-based DC-bus capacitor discharge strategies for the high-voltage permanent magnet synchronous motor (PMSM) powertrains used in EVs is discovered, which is mainly studied afterwards in this research.

(ii) To **propose a new winding-based DC-bus capacitor discharge method for the high-voltage PMSM powertrains with a slightly large inertia and a relatively small system safe current to quickly discharge the DC-bus capacitor and safely reduce the motor speed without voltage surge**. Considering that the traditional large-d-axis current injection-based (LDACI) discharge strategy might not be able to achieve the best discharge process (five-second discharge) when it is applied to a PMSM drive with extreme parameters, a new winding-based discharge strategy based on piecewise *d*, *q*-axis currents is proposed. The new strategy has stronger discharge capacity, and it can reduce the bus voltage quickly. Besides, in order to analyse the mechanism of the winding-based DC-bus capacitor discharge strategies explicitly, an energy flow model (EFM) is established, which can also contribute to designing a new winding-based discharge strategy.

(iii) To **develop a hybrid discharge strategy which employs the machine windings and external bleeder circuits to achieve five-second discharge in minimum sacrifice of bleeder size and weight for the high-voltage PMSM powertrains with highly extreme parameters**. Considering that the previous winding-based discharge strategies cannot achieve the best discharge process when the parameters of EV powertrains are highly extreme, a hybrid discharge strategy is developed. Both the external bleeding resistor (BR) and the discharge algorithms are designed. Actually, the new method is applicable to any EV powertrains regardless of the system parameters.

(iv) To **develop reliable winding-based DC-bus capacitor discharge strategies with fault tolerance against position signal loss**. The implementation of the discharge methods that rely on machine windings must use accurate position of a PMSM, which is usually provided by an external position sensor. Considering that the transmission wires of position signals are inclined to come loose or fall out during crash, which will influence the discharge process, the fault-tolerant algorithms are incorporated into the winding-based discharge strategies to improve the reliability of the system. In addition to the direct current regulation-based discharge algorithms, novel indirect current regulation-based discharge algorithms (by means of voltage and speed regulation) are proposed to enrich the winding-based discharge theories. The proposed fault-tolerant discharge methods contribute to improving the safety level of EVs in the crash conditions.

(v) To **develop parametric selection rules for the winding-based DC-bus capacitor discharge techniques for particular EV drives at the stage of design**. The pure winging-based discharge methods are parameter-dependent. They are not applicable to all EV powertrains especially when five-second discharge is requested. Therefore, it is necessary to pre-evaluate whether or which winding-based discharge technique is available when designing the discharge system for an EV. On the basis of theoretical parametric analysis, general principles that are able to select an effective and efficient discharge technique for a particular EV powertrain are developed.

1.2 Dissertation Outline

The dissertation consists of seven chapters. An outline of the structure is organized as follows:

This chapter introduces the motivation of this study. Besides, the main objectives and a brief introduction of the research contents are illustrated.

Chapter 2 reviews five types of injury hazards that the occupants might suffer during crash. For these injury hazards, the corresponding international and national crash regulations and the up-to-date practical technologies used for hazard prevention are presented. Importantly, the crash safety of the high-voltage powertrain-based EVs is addressed, and a valuable research topic is discovered through the review work.

Chapter 3 designs a new winding-based discharge technique for the PMSM powertrains with slightly large inertia and relatively small system safe current. Before designing the new method, an EFM is established to analyze the mechanism and characteristics of a winding-based discharge method. A comparative study is conducted to illustrate the stronger discharge capacity of the proposed winding-based discharge method.

Chapter 4 designs a hybrid discharge method relying on both machine windings and external bleeder circuits for the PMSM powertrains with highly extreme parameters. After discussing the defects of the traditional external bleeder-based strategy, the machine windings are adopted as the auxiliary plant for the external bleeder circuits, achieving a relatively lightweight and cost-effective discharge technique suited to any EV drives. The systematic design procedures for achieving the hybrid discharge approach are presented, which include two main sequential parts: external BR calculation and discharge algorithm design.

Chapter 5 designs fault-tolerant winding-based discharge techniques to improve the reliability of the discharge process. An enhanced second order sliding mode position observer and an adaptive sine-wave-based position observer are designed for the high-speed range and low-speed range operations, respectively. Meanwhile, in order to further enrich the winding-based discharge theories, in addition to the direct current control strategies, indirect current control strategies relying on voltage and speed regulation are introduced.

Chapter 6 discusses the general principles for selecting an effective and efficient discharge technique relying on the machine windings for a particular EV. The pure winding-based discharge strategies are discussed based on the following categories: instant non-zero d-axis, zero q-axis (NDZQ) current injection method, long-cycle NDZQ current injection method and long-cycle non-zero d-axis, non-zero q-axis (NDNQ) current injection method. By means of parametric analysis, the winding-based discharge strategy selection rules and their implementation procedures are presented.

Chapter 7 is the conclusion part. The main contributions of this study are summarized. Besides, the future work that can be carried out to further improve the crash safety of EVs is introduced.

Chapter 2
Review of EV Safety in Crash Conditions

Usually, a high-quality review is able to not only broaden the horizons of readers but also provide guidelines for the future study. However, there is little review work concerning the safety of the high-voltage powertrain-based EVs in the crash conditions now. In order to solve the problem, this chapter reviews the crash safety-related topics of the high-voltage powertrain-based EVs from the following two perspectives. (1) Possible injury hazards that the occupants are subjected to: The injury hazards in the crash conditions are categorized into five groups in this research, that is, physical injuries, electric shock, corrosion, intoxication and burn. The rationale behind these dangers is analyzed simultaneously. (2) Efforts made to prevent injuries: Firstly, the international and national crash regulations that are enacted to deal with those injury hazards are reviewed. It deserves to be mentioned that the regulations are mainly mandated to standardize the design, manufacturing and testing processes of automakers. Secondly, the up-to-date practical technologies that are adopted to reduce the aforementioned injury hazards are presented. Through the review work, the author discovered a valuable research topic that can be summarized as electric shock prevention by developing winding-based DC-bus capacitor discharge strategies, laying the foundations for the rest of this research.

2.1 Injury Hazards to Occupants During Crash

2.1.1 Physical Hazards

In theory, the passengers in an EV will confront the same physical hazards to the ones in the conventional ICE cars when crash emerges [15]. However, there are even few relevant studies that comprehensively introduce the physical injury hazards concerning the ICE cars during crash. In order to make it explicit for the ordinary

C. Gong, *Crash Safety of High-Voltage Powertrain Based Electric Vehicles*,
Springer Theses, https://doi.org/10.1007/978-3-030-88979-1_2

people to recognize the various injuries that they might suffer during crash, the physical injury hazards are analyzed in this section.

The forms of physical injuries include strain, sprain, bump, scratch, bleeding, fracture, crush and stab, etc. Intuitively, uncontrollable body movement, invasion of the external objects and intrusion of metal frame are the main factors causing these injuries.

On the one hand, when an EV crashes or hit by another car, not only the car body but also the human body will experience a fierce deceleration or acceleration process due to the external impact. The sudden body movements are very dangerous. Firstly, taking the restraint effects of the seat belt (serves in frontal and side collision) and seat (serves in rear collision) into account, the movements of the lower body, upper body and head, which result from the deceleration and acceleration processes, are different [16]. These lead to different tensile and compressive stresses between different parts of the occupant's body, and between the restraint devices and the passenger's body. However, one human body is only able to bear a certain limit of stresses, and once the limit is exceeded, the physical damage will inevitably occur [17]. For example, when an EV suffers a severe frontal collision when running with an extremely high speed, the occupants in the compartment witnesses a large deceleration. In this case, the seat belt is inclined to strain the chest and abdomen so as to injuries. In a more severe case, the passengers will suffer thoracic and lumbar rib and spine fracture. Besides, because the head displacement is much greater than the movement of the chest, the passengers might get a sprain on their vulnerable necks and spines. Secondly, the occupants might hit forcefully against the hard components inside the vehicle when leaning forward or backward during crash, causing the fatal visible wounds, such as bump, scratch, bleeding and fracture, and so on. For instance, it is possible that the head and legs of the occupant in the passenger seat run into the instrument board (ID) in a frontal impact case. At the moment, if the seat belt is properly used, only the above-mentioned injuries are predicted to be seen, but if the protective system does not work, it is possible for the occupant to hit the windshield (WS) or be thrown out of the car. In terms of the driver, apart from the injuries on the head and legs, the chest will also be in a pretty dangerous situation because of the compression from the steering wheel (SWL) (see Fig. 2.1) [18]. As for the side impact and rear impact scenarios, the passenger's head is prone to strike the door, B pillar and the seat back/headrest, respectively.

On the other hand, the vehicle body would get deformed and even fractured after a severe collision, which is dangerous as well. Firstly, as for the EV body shown in Fig. 2.2, it can generally be divided into three parts: frontal crumple zone, safe zone and rear crumple zone [19]. When a severe frontal or rear collision occurs, the bumper fascia, bumper beam, side beam and longitudinal beam will collapse so as to absorb the impact energy and spread the impact force all over the car frame, protecting the occupants and rechargeable energy storage system (RESS) in the safe zone from damage. However, the scattered pieces of the frame from the crumple zones fly uncontrollably and unpredictably, leading to the fact that together with the object dashed against the car, they might invade into the compartment and then hurt the occupants. In this case, the passengers will suffer from bump, scratch, bleeding,

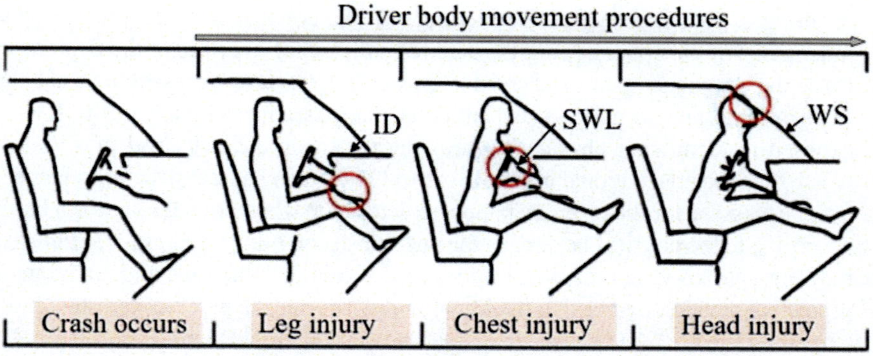

Fig. 2.1 Injuries on legs, chest and head for the driver in the frontal collision cases

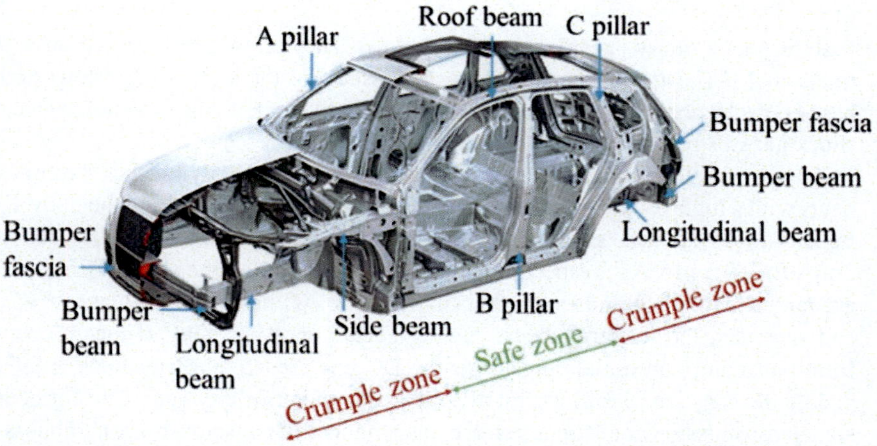

Fig. 2.2 Structure of the EV body and regionalism with respect to frontal and rear impact

fracture, crush and stab injuries. It deserves to be mentioned that it is difficult to take precautions to prevent the invasion of the external objects due to the chaos. Secondly, the road traffic injury and death rates increase as the intrusion of metal frame rises [20]. For example, the footwell intrusion induced by the frontal impact could cause severe injuries (e.g., fracture, crush and stab) on the lower body. What's worse, compared to the frontal and rear collisions, in which most of the impact energy can be absorbed by the specially-designed anti-collision zones, the impact force is directly imposed on the doors and B pillar in a side collision situation. Consequently, car deformation is unavoidable in this case once the external impact force is large. Considering the frame deformation in the side impact process, the occupants are facing the following hazards: (1) Hard and protruding geometric interior parts (such as the door armrests) can cause scratch and stab to the occupants' abdomens; (2) The intrusion of the door inner plate under the beam will cause vertical load on the pelvis, resulting in shear fracture of the sacrum; (3) The collapse and deformation of the inner panel of the upper door of the beam and the upper structure of the B-pillar will cause the occupants' head and chest to be swelled, fractured, crushed and stabbed.

It is important to address that although the passengers seated in the EVs are exposed to significant physical hazards during crash, the risk level of EVs is no worse than the conventional cars with respect to the mechanical structures. Hence, it is not necessary for the ordinary people to determine whether an EV is worth buying according to the physical hazards in the post-crash conditions. But as for automakers, they are spurred to continuously optimize the vehicle structures and elaborate the protective devices, decreasing the injury and death rates in a vehicle crash.

2.1.2 Electric Shock Hazards

Electric shock injury hazards are pretty special. They are seldom focused on in the traditional ICE cars, but they are important issues for the high-voltage powertrain-based EVs. Hence, for the sake of better understanding of EV crash safety, the electric shock hazards require to be discussed emphatically.

Intuitively, electric shock means that part of the human body touches the components with a high voltage, and a relatively large current passes through the body. IEC 62368-1:2018 defines "safe voltage" and "safe current" as 60 V DC, 30 V AC and 2 mA DC, 0.5 mA AC, respectively [21]. The reasons why the safe DC thresholds are larger than the AC ones are that firstly, 30 V AC has a peak voltage of 42 V, and secondly, the 'electric shock' feeling occurs on every peak of the AC waveform but mainly on initial contact for the DC one. Severe electric shock is lethal, and the reasons are as follows. Firstly, the inside bioelectricity generated for neural signal transmission could be interfered, causing loss of consciousness, numbness or tingling. Secondly, the abnormal stimulation can make the muscles and organs fail to work normally. For example, the symptoms of muscle spasms, breathing problems, vision or hearing problems and irregular heartbeat (even worse, heart stopping) appear. Thirdly, as a result of the thermal effect of current, body burns would occur, destroying the physiology of victims.

Nowadays, for the sake of high-power density and high efficiency, high-voltage propulsion systems are widely used in EVs [22, 23]. For example, the nominal DC-bus voltages of Renault Zoe, Tesla Model 3, BYD Tang and Taycan are 346 V, 350 V, 640 V and 800 V (far beyond 60 V), respectively. Intrinsically, these high-voltage drives would pose potential electric shock and complication hazards to the passengers inside and nearby the cars, which is one of the main concerns from both manufacturers and customers. Concerning this problem, as for a normal EV (before crash), the electric shock risk prevention methods include physical isolation and extra precautions [22]. The physical isolation refers to insulation. Usually, insulated cables and insulated housing are employed to prevent a person from contacting the live parts directly. In terms of the extra precautions which include equipotential connection, ground connection, automatic monitoring and protection systems, they work when the insulation is not totally effective. The equipotential connection and ground connection can transfer the high voltage to the vehicle chassis or ground, ensuring that no large current flows through the human body. The automatic monitoring and protection systems not only have the alerting function but also isolate the source of high voltage, protecting personal safety from leakage of electricity in the equipment. Even varieties of measures have been employed to prevent electric shock risks, EVs are not one hundred percent safe in practice. What's worse, the electric shock risks become higher in the crash conditions.

There are several reasons why an accident would exacerbate the electrical safety problems. Firstly, the metal frame deformation and body fracture expose the high-voltage components to the position where a passenger or rescuer can contact directly (physical isolation is broken). As for a normal EV, to prevent direct contact with live parts endowed with high voltage but low energy, the live parts in the passenger or luggage compartments must be protected to a degree of at least ingress protection (IP) XXD, and enclosures in other areas must have a protection degree of at least IPXXB, where IPXXD means a wire with 1 mm in diameter and 100 mm long cannot contact the live parts thorough joint gaps and IPXXB means a (test) finger cannot contact the high-voltage components [23]. However, vehicle deformation is prone to widen the joint gaps, reducing the IP level. Once the thin fingers of a person are able to be stick into the gaps near the live parts, the potential electric shock risks increase. Secondly, car deformation may lead the metal frame and chassis to contact the battery electrode or the joints between battery packs and power electronic devices (precautions lose efficacy). This issue arises from the RESS system of EVs. Generally, the positive and negative electrodes of the battery packs are connected with the power devices (e.g., DC/DC converter and DC/AC inverter) through binding posts. The insulation covering the terminals might be damaged or fall off due to the impact force. At the moment, it gets possible for the fractured frame or chassis to touch the binding posts, transferring the battery voltage to the bodywork. Once the occupants or rescuers touch a door, electric shock injury occurs. Thirdly, insulation failures might occur to the DC- or AC-bus cables during crash. In this case, on the one hand, without electrical isolation measures, both the positive and negative sides of the battery could come into contact with the bodywork through the damaged cables when encountering insulation failures. Similar as the second reason, the battery voltage becomes a threat to the safety of passengers in this situation. On the other hand, even though there are electrical isolation measures and they work effectively, the residual energy stored in the DC-bus capacitor and propulsion machine could also support the high-voltage bodywork [24] through the leaky cables.

Overall, the human body cannot withstand large voltages and currents. However, the high-voltage powertrains of a post-crash EV might trap the occupants in a compartment where they are inclined to suffer electric shock. Undoubtedly, this is dangerous. This is a typical feature of EVs by contrast with the conventional ICE cars.

2.1.3 Corrosion, Intoxication and Burn Hazards

Corrosion, intoxication and burn refer to the RESS-related injuries in this part. The traction batteries of EVs will pose potential risks to the occupants during crash, and this is drawing wide attention throughout the world.

As for the most commonly-used lithium-ion batteries, the internal materials are shown in Table 2.1 [25]. It can be seen that the power batteries are rich in the corrosive, volatile, toxic and flammable substances, especially in the electrolyte. In the normal cases, the materials inside the batteries are not harmful because they are stably enclosed in shells. But in an accident, the battery packs are inclined to suffer severe squeezing, impact and bending, with their structures damaged. Then, many hazards will arise.

When the impact force is imposed on the battery packs, they would deform and even rupture. Besides, the fractured metal frame is able to pierce the shells. In these cases, the electrolyte leaks out and evaporates [26, 27]. First of all, when an occupant's skin contacts the electrolyte, it will corrode. Secondly, the evaporative gas mixes with air to decompose and then (or directly) generate new corrosive and irritating gases that can be combined with water vapor, surrounding the passengers. Unluckily, once the gases enter the human body, the organs (e.g., eyes, skin, digestive tract and respiratory tract, etc.) will suffer corrosion. Thirdly, in addition to corrosion, both the electrolyte and the new gases are usually toxic. Long-term exposure to them could damage the nervous system, causing headache, dizziness, fatigue and nausea, etc. A typical solvent that satisfies the above-three features are $LiPF_6$. Both itself and its decomposition product PF_5 are highly corrosive and toxic.

Apart from electrolyte leakage, thermal runaway of batteries is another common phenomenon in crash [28, 29]. Impact, vibration, deformation and pierce could destruct the separators in the battery, leading to short circuit between the positive and negative electrodes. Then, a large amount of heat is generated within a short while, and the battery starts to burn. Meanwhile, as the temperature of battery increases sharply, the electrolyte evaporates and decomposes quickly [30]. Unluckily, the evaporative gases are flammable as well (such as, PF_5), facilitating the fire to spread rapidly. Even worse, when the concentration of the flammable gas is high enough, it explodes. As for the occupants seated in the EV, fatal burns and wounds could arise from the battery fire and explosion. Additionally, because the combustion products

Table 2.1 Internal materials of lithium-ion batteries

Components	Materials
Anode	$LiCoO_2$, $LiMn_2O_4$, $LiFePO_4$, $LiNiCoMnO_2$
Cathode	Carbon materials, like graphite and carbon fiber
Separator	Polypropylene, polyethylene, composite membrane
Electrolyte	(1) Lithium salts: $LiClO_4$, $LiPF_6$, $LiBF_4$, $LiAsF_6$. (2) Solvent: $C_3H_4O_3$, $LiPF_6$, $C_4H_6O_3$, PF_5, $C_5H_{10}O_3$, $C_3H_6O_3$, $C_4H_8O_3$. (3) Additives

of the electrolyte usually include CO, HCl, HF, HCN, C_6H_6, C_7H_8, the poisoning hazard is unavoidable in practice.

2.2 Regulatory Activities Concerning Crash

Vehicle crash regulations are requirements that automobiles must satisfy in order to be approved for sale in a particular country or region. Car manufacturers have to design products that assure the respective safety requirements because the regulations are usually mandated by legislation and administered by a government body. The worldwide crash regulations can be categorized into two groups [13, 31–34]: the international ones and national ones. The international regulations refer to United Nations (UN) Global Technical Regulations administered by UN Economic Commission for Europe (ECE), which is implemented in Europe. The national regulations include Federal Motor Vehicle Safety Standards (FMVSS) in the United States (US), China Guobiao (GB, GB/T) Standards, Korea Motor Vehicle Safety Standards (KMVSS), Japan Test Requirements and Instructions for Automobile Standards, Australian Design Rules (ADR) and India Automotive Industry Standards (AIS), etc. It needs to be mentioned that the regulations administered in China, Japan, Australia and India are mainly amended with reference to the UN ECE series, while South Korea has referenced the US series [35–37]. Considering that the derived national regulations inherit the safety considerations of the corresponding primary series, this section will mainly introduce the provisions in the UN ECE regulations and US FMVSS regulations.

2.2.1 Regulations Concerning Physical Hazards

Table 2.2 lists the specific regulation files (both international and national ones) with regards to occupant protection against physical injuries in different vehicle collision situations. These regulations set clear requirements for the maximum acceptable damage to the occupants and car body under specified/test collision conditions.

(a) **Frontal impact**

UN ECE R94 and R137 are developed for the 40%-overlap frontal impact and 100%-overlap frontal impact tests, respectively. According to [38], when a vehicle hits against a rigid barrier with a 40% overlap at the speed of 56 km/h, the following requirements need to be satisfied:

- Head injury criteria (HIC_{36}) ≤ 1000, and the resultant head acceleration (RHA) shall not exceed 80 g for more than 3 ms.
- Neck injury criteria (NIC) shall not exceed neck tension and shear criterion in [38].

Table 2.2 International and national regulations with regard to occupant protection against physical injuries

Collision	Areas						
	UN (Europe)	US	China	South Korea	Japan	Australia	India
Frontal impact	UN ECE R94, R137	FMVSS 208	GB/T 20913, GB 11551	KMVSS 102	Article 18, Attachment 23	ADR 73/00, 69/00	AIS 098, 096
Side impact	UN ECE R95	FMVSS 214	GB 20071	KMVSS 102	Article 18, Attachment 24	ADR 72/00	AIS 099
Side pole impact	UN ECE R135	FMVSS 214	GB 37337	KMVSS 102-4	Article 18	ADR 85/00	IS 12009
Rear impact	UN ECE R32	FMVSS 202a	GB 20072	KMVSS 99	Article 18, Attachment 34		AIS 101

- Neck bending moment (NBM) $\leq$ 57 N m.
- Thorax compression criterion (ThCC) $\leq$ 50 mm.
- Viscous criterion for the thorax ($V \cdot C$) $\leq$ 1.0 m/s.
- Femur force criterion (FFC) shall not exceed force-time performance criterion (curve) in [38].
- Tibia compression force criterion (TCFC) $\leq$ 8 kN.
- Tibia index (TI) $\leq$ 1.3.
- Movement of the sliding knee joints (MSKJ) $\leq$ 15 mm.
- Residual steering wheel displacement (RSWD) $\leq$ 80 mm in the upwards vertical direction and 100 mm in the rearward horizontal direction.
- No door shall open during crash.
- No locking of the locking systems of the front doors shall occur.
- Without using tools, open at least one door and remove the intact dummies (occupants) from the vehicle.

Comparatively speaking, apart from the test conditions (full overlap, crash speed = 50 km/h), the requirements in UN ECE R137 are slightly different. First, the NIC and FFC are given specifically, which are,

- Axial neck tensile force (NTF) $\leq$ 3.3 kN.
- Fore/aft neck shear forces (NSF) $\leq$ 3.1 kN.
- FFC $\leq$ 9.07 kN.

Second, the ThCC becomes lower:

- ThCC $\leq$ 42 mm.

When it comes to the US standards, FMVSS 208 covers both the offset- and full-frontal crash scenes. According to [39], as long as the crash speed is not higher than

48 km/h and the crash angle is smaller than 30°, the following occupant injury and vehicle damage criteria should be satisfied:

- $HIC_{36} \leq 1000$, $HIC15 \leq 700$, and RHA shall not exceed 60 g for more than 3 ms.
- Axial TNF $\leq$ 4.17 kN.
- Axial neck compression force (NCF) $\leq$ 4 kN.
- $N_{ij} \leq 1$ at any time during crash, where N_{ij} represents a comprehensive factor related to shear force, axial force and bending moment.
- Compressive deflection of the sternum relative to the spine shall not exceed 76 mm.
- Force transmitted axially through each upper leg shall not exceed 10 kN.
- All portions of dummies (occupants) shall be contained within the compartment.

It can be noted that there are obvious differences in both safety considerations and critical values between the UN and US frontal impact regulations. However, neck, thorax and legs are focused on in both series, representing that these body parts are more vulnerable during crash. Additionally, the regulations require the vehicle body to be strong and robust enough to protect the occupants from physical injuries as far as possible.

(b) **Side impact**

In UN ECE R95, the crash condition is defined as the case that a moving deformable barrier (MDB) hits against the vehicle side at the speed of 50 km/h. During crash, the performance criteria are summarized as follows:

- $HIC_{36} \leq 1000$ when there is head contact.
- Thorax rib deflection criterion (RDC) $\leq$ 42 mm.
- $V \cdot C \leq 1.0$ m/s.
- Pelvis pubic symphysis peak force (PSPF) $\leq$ 6 kN.
- Abdominal Peak Force (APF) $\leq$ 2.5 kN.
- No door shall open during crash.
- Without using tools, open doors and remove the intact dummies (occupants) from the vehicle.
- No interior device or component shall get detached to become sharp projections or jagged edges.

In terms of the crash conditions in FMVSS 214, the vehicle is struck on either side by an MDB at the speed of 53 km/h. During crash, the up-to-date requirements that need to be satisfied include:

- $HIC_{36} \leq 1000$.
- Thorax RDC $\leq$ 44 mm.
- Pelvis PSPF $\leq$ 6 kN.
- Sum of the front, middle and rear abdominal forces $\leq$ 2.5 kN.
- No door that is struck shall separate totally.
- No door that is not struck shall disengage from the latched position.

As for the side impact regulations, the safety considerations (head, thorax, pelvis, abdomen and door) in the two series of regulations are the same, and the criteria for them are very similar. In addition, the doors of the vehicle during crash attract much more attention than the other components.

(c) **Side pole impact**

As for UN ECE R135, the vehicle speed at the moment of first vehicle-to-pole contact is 32 km/h. The injury criteria are:

- $HIC_{36} \leq 1000$.
- Peak lateral shoulder force (LSF) $\leq$ 3 kN.
- Thorax RDC $\leq$ 55 mm.
- Pelvis PSPF $\leq$ 3.36 kN.
- Abdominal RDC $\leq$ 65 mm.
- Resultant lower spine acceleration (RLSA) $\leq$ 75 g.
- No door that is struck shall separate totally.
- Doors that are not struck shall remain latched.

FMVSS 214 adopts the same crash speed to UN ECE R135, while the requirements that need to be meet are slightly different:

- $HIC_{36} \leq 1000$.
- Thorax RDC $\leq$ 44 mm.
- Pelvis PSPF $\leq$ 6 kN.
- Sum of the front, middle and rear abdominal forces $\leq$ 2.5 kN.
- No door that is struck shall separate totally.
- Doors that are not struck shall remain latched.

Interestingly, the criteria for side pole impact are the same as those for side impact in FMVSS 214, while the provisions in UN ECE R135 show noticeable differences compared to those in UN ECE R95. The allowable damage to passenger head, thorax, pelvis, abdomen and vehicle doors are still focused on in the side pole impact scenes.

(d) **Rear Impact**

In UN ECE R32, the velocity of collision is between 35 and 38 km/h and the moving barrier is rigid. As for UN ECE R32, the requirements are mainly set for the vehicle body:

- Lengthwise displacement $\leq$ 75 mm.
- No rigid component in the passenger compartment shall constitute a risk of injury to the occupants.
- No door shall open during crash.
- Without using tools, open doors and remove the intact dummies (occupants) from the vehicle.

In FMVSS 202a, the crash speed is set as 17.3 km/h and the following injury criteria are specified:

- Posterior angular rotation (PAR) between the head and torso $\leq 12°$.
- $HIC_{15} \leq 500$.

As far as rear crash is concerned, UN regulations mainly focus on the vehicle body conditions while the head injuries are taken into account in the US regulations.

2.2.2 Regulations Concerning Electrical Hazards

Electrical safety regulations are developed to avoid or reduce the high-voltage electric shock and RESS-related hazards during crash. Unlike the regulations in Table 2.2, which can be also applied to the traditional ICE cars, the electrical safety regulations are only applicable to EVs. Now, the requirements in one particular international or national crash regulation file are usually shared between the frontal, side and rear impact cases [40]. For example, although the electrical safety regulations are only specified in UN ECE R94 (frontal impact), they are also applicable to the side impact and rear impact situations for the approved EVs in the European market. Besides, FMVSS 305, Attachment 111, GB/T 31489 and Article 91 are the electrical safety regulations in US, Japan, China and South Korea, respectively.

With regards to the protection against electrical hazards following the impact test, the amendments specify five main performance criteria in UN ECE R94:

- Within 30 min, the liquid that spills outside the passenger compartment ≤ 5 L, and no liquid shall spill into the passenger compartment.
- RESS shall remain in the installed location and RESS components shall remain inside RESS.
- Insulation resistance between the bus and the chassis ≥ 500 Ω/V for AC bus or 100 Ω/V for DC bus.
- Bus voltage decreases under 30 VAC or 60 VDC fast or energy on bus decreases under 0.2 J fast (5 s is the best and 60 s is the worst).
- Have a protection degree of IPXXB and resistance between exposed conductive parts and electrical chassis < 0.1 Ω.

It deserves to be mentioned that the provisions in FMVSS 305 are nearly identical to those in UN ECE R94. This means that most of regions in the world have reached a consensus on the electrical safety considerations and criteria.

2.2.3 Discussion and Future Challenges About Regulations

The crash regulations are mainly developed for test, and as long as the EVs satisfy the requirements, they will be locally approved for sale. Therefore, the vehicle manufacturers take on the main responsibility for making their products qualified. As an ordinary person, it is needed to recognize that the EVs that we can buy or see in the daily life are pretty safe in the crash conditions because they must satisfy the relative legislations.

Nevertheless, the current regulations still require to be improved considering the following aspects:

(1) Regulations in some regions are not comprehensive. For example, the rear impact regulations in Australia need to be established in allusion to the physical hazards. Besides, many countries still lack uniform electrical safety regulations, such as India. Hence, it is urgent to determine the relevant regulations as soon as possible through introduction or establishment so as to adapt to the rapid development of EVs.
(2) Regulations need to cover more aspects because many residual risks still remain [40]. For instance, Firstly, the amendments specify requirements to control the movement of a RESS, but there are no requirements for its structural integrity that is closely related to short circuit, rupture, fire and explosion hazards. Secondly, the amount of electrolyte that leaks might increase if an EV rolls over following a collision. In this case, the regulations about electrolyte spillage deserve modifications.
(3) Coordination of various national regulations is of great significance. Specifically, although the national regulations can be different, this might bring about challenges for vehicle manufacturers to sell their products in other regions.

2.3 Technologies for Reducing Injury Hazards to Occupants After EV Crashes

Technical supports are indispensable for ensuring an EV to satisfy the crash regulations and preventing the occupants from various injuries (e.g., physical injuries, electric shock and corrosion injuries, etc.) in the crash conditions. This part mainly introduces the current commonly-used technologies that contribute to occupant injury hazard reduction.

2.3.1 Technologies for Reducing Physical Hazards

According to the aforementioned analysis of the reasons why the occupants get physically hurt during crash, to suppress vehicle intrusion and to suppress direct human body impact should be the two most crucial targets for the purpose of injury prevention, which mostly relies on the anti-collision structures and passenger restraint system of a vehicle, respectively. As in Table 2.3, the anti-collision structures of an EV mainly comprise bumper fascia, bumper beam, longitudinal beam, the rest of car frames (A, B, C pillars, side beam and roof beam, etc.) and doors. In terms of the passenger restraint system, it consists of seatbelts, airbags and safety seats. Moreover, considering that once the passengers hit a sharp interior part of an EV,

Table 2.3 Technologies for reducing physical injury hazards

Injury hazards	Targets	Technical aspects	References
Physical injuries	Vehicle body intrusion suppression	Bumper fascia	[42, 43]
		Bumper beam	[44–47]
		Longitudinal beam	[48–53]
		Rest of car frames	[54–60]
		Doors	[61–63]
	Human body impact suppression	Seatbelt	[64–67]
		Airbag	[68–72]
		Safety seat	[73]
		Interior trims	[41, 74, 75]

the impact force imposed on the human body will become significantly large. Therefore, in order to optimize the vehicle interior trims should be treated as an important measure for suppressing human body impact [41]. In Fig. 2.2, the bumper fascia serves as a buffer, absorbing the impact energy foremost during crash [42, 43]. When the collision is slight, the bumper fascia is effective for preventing the impact force from being transmitted to the other parts of the vehicle body. But if a severe collision occurs, it will collapse immediately. In order to enhance the energy absorption ability of the bumper fascia, material and structure optimization is necessary. Specifically, [42] employs the carbon fiber reinforcement vinyl ester with microsphere (CRVeM) material to replace the traditional polypropylene plastic. In [43], a honeycomb structure which is obtained by a series and systematic arrangement of thin edged hexagonal cells is designed to improve the overall performance of bumper fascia. Because the honeycomb structure has high compression and shear properties in the axial direction, it is much stiffer and stronger than the traditional structure when confronting a frontal/rear collision. After the bumper fascia loses efficacy in a severe collision, the bumper beam, which is also known as anti-collision beam, directly bears the external impact force. Similarly, when designing and optimizing the anti-collision beam, material and structure are the main considerations for enhancing the protective capability [44–47]. An aluminum alloy sandwich panel beam is proposed in [44], of which structure consists of three layers. The first and third layers are made of metal and metal alloy, or composite materials such as glass fiber and high-pressure crushing board, etc. The energy absorption capacity is analyzed by the drop weight test, and it is found that the upper panel supported by the core plate and the lower plate is more resistant to penetration. Literature [45] and [46] develop composite bumper beam via the design optimization process combined with the impact analysis. The composite is devised to substitute for the conventional steel and glass mat thermoplastic (GMT) for reducing the weight but increasing the energy absorbing capacity of bumper beam. In [47], a novel composite bumper beam is developed to improve energy absorption and lighten the weight of bumper by the means of Fruit Fly Optimization Algorithm (FOA). It deserves to be mentioned that [47] compares

HIC_{36} before and after adopting the composite material experimentally, proving that the new material is able to reduce HIC_{36} by around 6.37%.

Longitudinal beam is connected with the bumper beam, and it absorbs most of the impact energy in a severe crash. There are two reasons why the longitudinal beam has such a powerful ability. Most importantly, as parts of the longitudinal beam, crash boxes installed between the bumper beam and the longitudinal beam are specially designed for impact energy absorption. The crash boxes will crumple during crash so as to reduce the impact force that can continue to travel backwards. Secondly, the longitudinal beam could bend spontaneously. At present, the most common method for exerting the function of a longitudinal beam is structure optimization. In [48], a new beam structure with variable rolled blank and variable cross-sectional shape is proposed, which is proven to be more crashworthy theoretically. Literature [49] develops a multi-cell thin-walled structure, and based on parametric analysis, the geometric parameters of the new structure are designed. In addition, [50] specially studies the impacts of the thickness on the anti-collision performance. By using a new surrogate model and non-dominated sorting in genetic algorithms-II (NSGA-II), the optimal results are obtained. The structure of the crash box is focused on in [51–53], and a square crash box with holes, a cutting-the-screw-thread structure and a multiple-cavity structure are designed in these researches, respectively.

Compared to bumper fascia, bumper beam and longitudinal beam, the rest of the car frames as well as the doors are more inclined to intrude into the EV compartment and cause injuries because they are nearer to the occupants. The strength and stiffness of these parts can be improved by optimizing the material, structure and processing techniques (technics). In terms of the car frames, [54] employs a carbon-fiber-reinforced plastic (CFRP) to substitute the conventional steel materials for reducing the weight but increasing the crashworthiness of the pillars. Simultaneously, the shape, stacking sequence and thickness of the pillars are optimized to minimize the deflection profile in the study. Literature [55] optimizes the B pillar using new innovative Tailor Rolled Blank (TRB) structure. In [56], the strength-ductility synergy of dual-phase (DP) steels with high volume fraction of martensite and various tough structures are obtained to improve the strength of the vehicle beams. Moreover, bionics have been used for optimizing the crashworthiness for overall car frame. For example, a turtle shell structure is proposed in [57], of which mechanical characteristics are analyzed in the research. Considering that the welded junctions between different parts (e.g., beams and pillars) of the car frame are so weak that it is easier to fracture during crash, the continuous laser welding techniques endowed with better flatness and smaller gaps are used to replace the traditional spot welding with poor continuity [58–60], increasing the strength of the junctions. As far as the doors that are responsible for absorbing most of the kinetic energy in side impact are concerned, door beam is usually embedded for structure reinforcement. Literature [61] and [62] study the material, geometry and connection of the side door beams to reduce damage and reaches the following two conclusions. Firstly, any plastic displacement of the door plate should be avoided; Secondly, increasing the door beam thickness causes a peak in the beam rigidity and the crush forces, so there should be a reduction in the door beam deflections. By the means of finite element analysis tools, the strength of the door structure is analyzed and optimized in [63].

As the most crucial components of the passenger restraint system, seatbelts must be installed in every EV. The standard seat-belt system consists of three-point seat belt, a load limiter and a retractor pre-tensioner. However, [64] points out that there is still much room to improve the current seat-belt technologies when taking the belt slack issue into account. In other words, the standard seat-belt system is not optimal. Now, the widely studied technical aspects with regards to the seat belts include slack suppression, load distribution optimization and controllability. In [65], to reduce the slack in the shoulder portion and the lap portion, an anchor pre-tensioner and a buckle pre-tensioner are added and evaluated. Literature [66] uses finite element approach to optimize load distribution by adjusting the buckle and anchor positions, minimizing the compression force imposed on the passenger body during crash. In [67], an algebraic governor is employed to estimate and detect the crash events, and according to the degree of a crash, a linear feedback loop algorithm is implemented to control the behaviors of the seatbelts. It is found that compared to an uncontrolled seatbelt system, a controlled one is able to reduce the injury index by 15%.

Airbags are important to prevent the occupants from hitting against the hard components in a severe crash. The up-to-date airbag technologies used for injury hazard reduction mainly include four aspects: increasing the number of airbags, structure optimization, accurate deployment control and inflation process optimization. Firstly, in addition to the two compulsory frontal airbags that can only offer protection in a frontal collision, side airbags, curtain airbags and knee airbags, etc. (up to 24 airbags) have been employed in modern cars to comprehensively protect the occupants [68]. Secondly, scholars and engineers never stopping optimizing the geometric parameters and shape of the airbags. For example, the multi-objective genetic algorithms are employed to design the height and width of an airbag and the area of vent hole in [69]. Literature [70] designs a new tubular airbag to reduce the dosage of gas generation, and the simulation results show that the stiffness and protection performance of the new airbag are better than the traditional ones. Then, the deployment moment for airbags is determined conventionally by diversifying the type and location of crash-related sensors. Whereas, [71] argues that the existing crash algorithms are prone to make an erroneous judgment, missing the time frame for airbag deployment or generating a wrong deployment signal. On this ground an improved algorithm that combines both the pre-crash information and the crash-related-sensor signals is proposed to ensure the precision and reliability of airbag operations. Finally, when the airbags are triggered in crash, they must inflate quickly. Literature [72] reduces the ignition delay time (24%) of the airbag by using a boron/potassium nitrate/5-amino-1H-tetrazole-based pyrotechnic igniter.

Safety seats are developed to protect the child passengers considering that the seatbelts and airbags are not applicable to them. The categories, properties and reinforcement approaches are reviewed in [73]. According to the research, the structure (surface, connection, size and belt regulation, etc.) of the safety seat is attracting much attention currently for the sake of higher protective ability.

The materials of the interior trims should be soft, and sharp edges should be avoided. Only by using these technologies could the impact be reduced once the human body strikes them. Literature [41] and [74] adopt polymer/polyurethane sponge layers as the car interior trim materials. In [75], the A-pillar trim is optimized by using smooth surfaces.

2.3.2 Technologies for Reducing Electric Shock Hazards

The reasons and regulations indicate that the IP level, cable insulation, high (battery) voltage isolation and residual energy dissipation (DC-bus voltage reduction) require to be paid attention to for the sake of electric shock injury avoidance (see Table 2.4).

There are two factors influencing the IP level in an EV crash, namely, joint gaps and live parts. The joints gaps of the compartment widen due to the external impact force and vehicle body deformation. Hence, the overall car frame should be designed and optimized to be with high anti-collision and energy-absorbing capabilities. The technologies illustrated in [42–63] are effective for preventing joint gaps from increasing significantly (at least IPXXB). However, no matter how strong the car frame is, it is still possible that the joint gaps get large in an extremely severe crash. Considering this, another applicable solution is to place the live parts that include cables, converters, motors, compressors and inverters, etc., far away from the joint positions [76]. In addition, the enclosures of the live parts can be directly connected with the chassis, causing an equipotential effect [77]. In this situation, even if the occupants touch the live parts, the electric shock risks can be averted because no current can flow through the human body.

Insulation can be categorized as basic insulation, auxiliary insulation, double insulation and reinforced insulation [78–81]. Double insulation means a combined structure composed of the basic insulation and auxiliary insulation. Reinforced insulation has the similar protective ability as double insulation. Generally speaking, the conductive parts cannot be reached when applying double and reinforced insulation [82]. In other words, these two kinds of insulation are so robust that they are not easy to fail in a crash, avoiding the direct and indirect contact with the passengers. However, considering cost effectiveness, it is not affordable to use double/reinforced insulation everywhere, so they are only applied to part of the crucial components now.

Table 2.4 Technologies for reducing electric shock injury hazards

Injury hazards	Technical aspects	References
Electric shock	IP level	[76, 77]
	Insulation	[78–82]
	High voltage isolation	[83–88]
	Residual energy dissipation and DC-bus voltage reduction	[89–97]

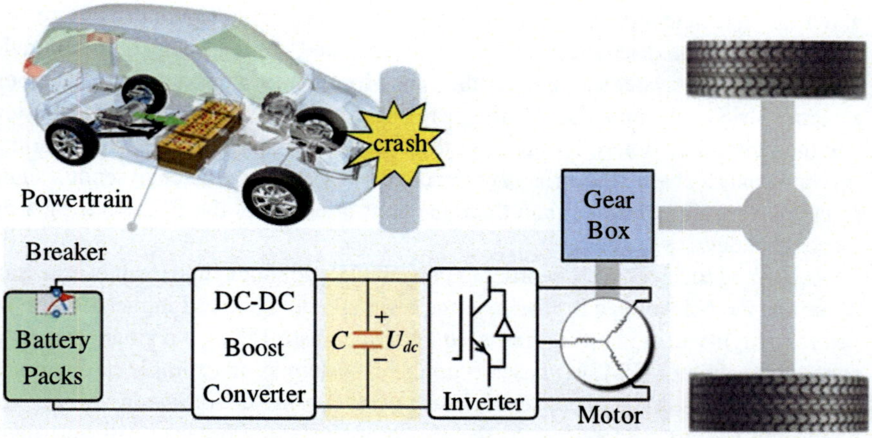

Fig. 2.3 Block diagram of the structure of the EV powertrain

Battery is the most principal source for producing a high voltage. If the battery can be disconnected from the transmission cables through a breaker embedded in or near the battery packs (see Fig. 2.3), which is called high voltage isolation, the electric shock injury hazards can be remarkably reduced regardless of the IP level reduction and insulation failures during crash. The disconnect devices used for high voltage isolation include electromechanical relays [83], semiconductor devices [84] and pyrotechnical disconnect devices [85, 86]. Electromechanical relays are susceptible to contact problems because of their design, which include corrosion, bridge-formation and arcing failures. The contacts might stick together over longer periods due to the high working currents and vibrations. As for the semiconductor devices, because the operating DC current is usually large, high voltage ($\mathrm{d}i/\mathrm{d}t$) might be generated and applied to the power devices, resulting in permanent damage. It is widely acknowledged that the pyrotechnically powered battery disconnection devices are reliable. As is shown in Fig. 2.4, when a triggering signal comes, the electrical initiator generates an initial spark, and fires the chemical transformation of the pyrotechnical charge. The rapid build-up of pressures in the combustor accelerates the piston, using the chisel built into the piston to cut off the conductor rail. Another interesting topic

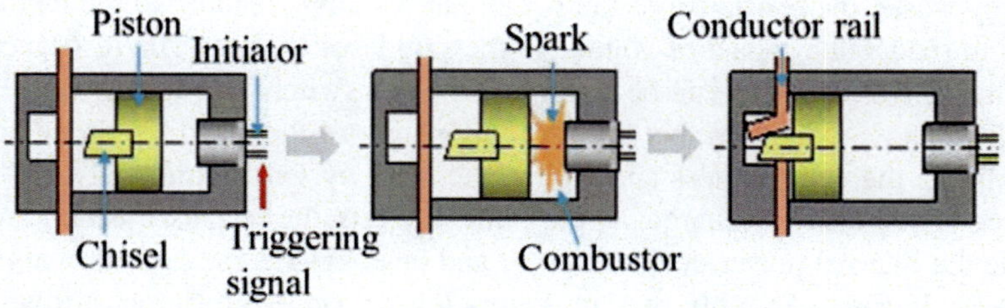

Fig. 2.4 Working mechanism of a pyrotechnical disconnect device [85]

about the disconnect devices is how to generate a triggering signal. Literature [87] introduces that the disconnect device can be activated via an airbag control module. Once the airbag system ascertains the fact of collision, the battery management system (BMS) will immediately trigger the actions. In [88], both the airbag system and the powertrain controller are used to generate a triggering signal. The working mechanism is that after the airbag system confirms a severe collision, it sends a signal to the powertrain controller, and then, the controller forces the BMS to trigger the disconnect devices.

According to Sect. 2.2.2, in order to prevent electric shock injuries, it is necessary to reduce the bus voltage or dissipate the residual energy stored in the powertrains fast for a high-voltage powertrain-based EV after crash. Taking a permanent magnet synchronous motor (PMSM) based propulsion system as an example (the motor in Fig. 2.3 is a PMSM), once a severe crash occurs, a protection mode will be triggered immediately. According to the above analysis, the breaker will be tripped immediately to isolate the battery from the other components. In addition, the axle is disconnected from the traction motor by the gear box and the PMSM just rotates with no load [89]. However, in this case, the residual electrical and mechanical energy stored in the capacitor and motor respectively will maintain the DC-bus (capacitor) voltage at the high level for long (several minutes) if no energy consumption means are adopted, violating the high-voltage safety requirement. Since the DC-bus capacitor voltage cannot go down to the safe level spontaneously, this provides strong driving force to develop proper DC-bus capacitor discharge techniques. So far, several discharge strategies have been developed, and they can be grouped into two categories: external bleeder-based method and internal power device-based method. (1) External bleeder-based discharge: External bleeder circuits that are installed in parallel with the DC-bus capacitor can be employed to reduce the bus voltage in emergency. The principal component of the bleeder is a bleeding resistor R_b that is able to consume the residual energy in the form of heat. According to whether a controllable switch s is adopted in the discharge circuit, the bleeder-based discharge can be further divided into passive discharge and active discharge (see Fig. 2.5a) [90–92]. The common characteristics for the two modes are that the transistors in the inverter are sealed off when a crash monitoring sensor detects emergency, and the energy stored in the machine will flow to the bus side through the uncontrolled rectifier composed of six free-wheeling diodes. As for passive discharge, the bleeding resistor (BR) always works even in the normal situations. Hence, to cut down on energy waste, the resistance of the BR should be large, leading to the fact that it is hard to reduce the capacitor voltage to the safe level quickly. The resistance of the BR for active discharge can be small because the switch is in the off-state unless a crash signal is diagnosed, but the size and weight of the bleeding circuit are huge, sacrificing the compactness and cost of the drive system significantly. (2) Internal power device-based discharge: As the name suggests, the external bleeders are absent while the internal power devices (motor and inverter) are used for discharge in this strategy. In Fig. 2.5b, only two transistors located on two different bridge arms of the inverter are switched on [93, 94]. The electrical energy stored in the capacitor is consumed by the two conducive transistors, and the mechanical energy in the motor

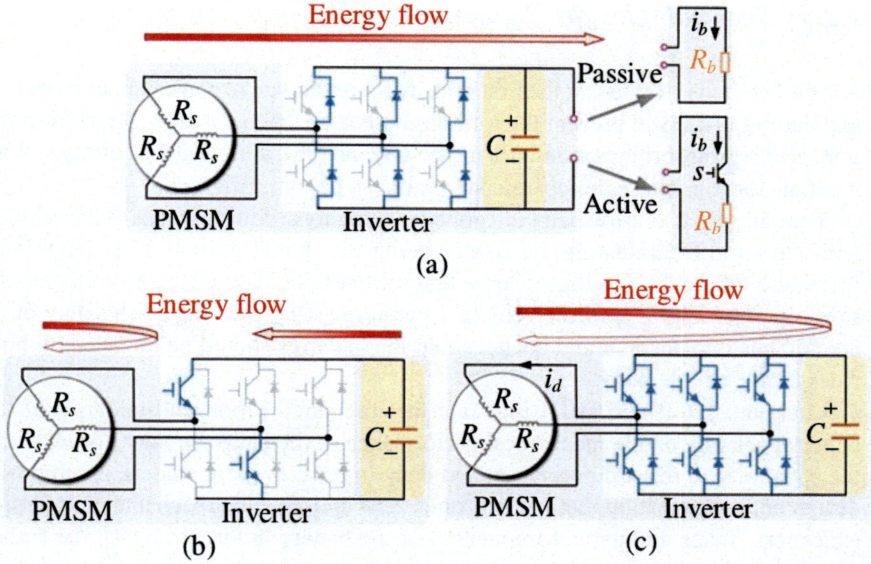

Fig. 2.5 Discharge techniques. **a** External bleeder-based discharge. **b** Short-circuit internal power device-based discharge. **c** Current injection-based discharge

is firstly converted into the electrical one and then consumed by both the transistors and the machine windings. One obvious disadvantage of this strategy is that due to the short-circuit effect, very large discharge current will be generated, posing risks to the health of the power electronics. Figure 2.5c illustrates the traditional large-*d*-axis current (*d*-axis current, together with *q*-axis current, α, β-axis currents in this study, is a concept in vector control for AC motors, which is illustrated in Appendices) injection-based (LDA-CI) discharge technique [89, 95–97], which directly functions the motor windings as the bleeder (thereby being known as winding-based discharge as well). By weakening the air-gap flux in the motor, the back electromotive force (EMF) that is tracked by the DC-bus capacitor voltage will get down to a pretty low level instantaneously. In this case, the residual energy is dissipated mainly by the machine windings through the controllable current in the form of heat loss. Comparatively speaking, when taking cost, size and safety into account, the machine winding-based discharge method by the means of current-injection operations has superior advantages over the external bleeder-based discharge and the inverter-based discharge approaches. Besides, without using extra circuits, the reliability of the discharge system increases.

2.3.3 Technologies for Reducing RESS-Related Hazards

As is illustrated in Table 2.5, the first priority for reducing the corrosion, intoxication and burn injuries is to prevent the lithium-ion battery from damage. Apart from the impact energy absorbing techniques in [42–63] and [98], aiming at the battery system itself, its locating place, holder and structure can be optimized.

The battery packs are usually integrated into highly reinforced areas of the vehicle, with the aim of eliminating the risk of being penetrated during crash conditions. Literature [99] and [100] demonstrate that the batteries should be located inside the safe zone (see Fig. 2.2) of the vehicle. Literature [101] takes the crash safety of an electric bus into account and declares that the batteries should be installed on both sides of the vehicle body.

Literature [102] and [103] illustrate that the stabilization and integrity of the battery pack rely on the mechanical performance of the battery holder, for instance, the capability to resist deformation and deny impact. In these studies, a composite design and optimization method is proposed to optimize the mechanical properties (thickness, shape and natural frequency) of the battery holder. In [101], the author found that the side around a battery box is the main collision zone, so it needs to be designed well. Meanwhile, this study analyses the energy absorption capacity, connection modes and stiffness of the side part of a battery box, and then, an improved battery box is developed in this research. Moreover, the high-strength material and reinforcing ribs are integrated into the battery box in [104] and [105], respectively. Patent [106] introduces the battery holder with outer and inner titanium alloy plates, which are of high anti-collision capacity.

To reinforce the battery pack/cell structure can effectively reduce the electrolyte leakage and large-scale fire problems [107]. In the study, the effect of strain rate behavior of the casing upon the cell average force-displacement response on lateral indentation, three-point bending, and axial compression load cases in investigated. Literature [108] has performed a punctured floor battery packaging by external indenter and damage on battery cell by indenter with and without offset is also simulated. Furthermore, experiments and simulation on the plasticity of the shell casing of battery cell have been performed in [109]. It was observed that considerable anisotropies exist on the shell due to rolling and deep drawing process and the experimental results are used to develop the plasticity and fracture model of the shell. There are several studies about battery module protection structure based on aluminum [110–112]. In [113], a lightweight protection structurer based on fiber metal laminate (FML) is designed to absorb the impact energy. Compared to the aluminum structure, FML consists of metal sheets bonded to fiber-reinforced plastic

Table 2.5 Technologies for reducing RESS-related injury hazards

RESS-related injuries	Battery placement	[98–101]
	Battery holder	[101–106]
	Battery pack/cell structure	[107–116]

layers, so it is of higher specific stiffness and strength. structural optimization of battery packs is focused on in [114–116]. By placing the energy-absorbing battery modules in the front of the vehicle, the optimum working pressure of the battery module can be fully used to effectively protect the driver.

Overall, according to the above analysis, it can be found that firstly, the material, structure and processing techniques are commonly optimized for the sake of high safety; Secondly, prevention of direct and indirect contact with the live parts and residual energy dissipation are the main aspects for the sake of electric shock risk reduction. Thirdly, it is important to strengthen the battery itself for the purpose of electrochemical and burn injury prevention.

2.4 Valuable Topic Requiring Further Study

As is illustrated in Sect. 2.3.2, in order to prevent electric shock injury hazards during crash, a winding-based discharge method can be employed to dissipate the residual energy stored in the high-voltage PMSM powertrain in an EV, reducing the DC-bus voltage to the safe level quickly. Considering that a winding-based discharge strategy is cost-effective, compact and high-reliability, it is extremely valuable in the engineering industry. However, few scholars and engineers focus on this competitive discharge method, and the relevant researches still remain in the preliminary application stage of the traditional LDA-CI scheme. There is still a long way to go before the theories and techniques about the winding-based discharge become mature. Consequently, the main objective of the rest of this research is to study the winding-based discharge strategies in depth with reference to different challenges encountered in the real applications, improving the crash safety of the high-voltage PMSM powertrain-based EVs.

2.5 Summary

This chapter reviews three aspects about EV safety topics in the crash conditions, namely, the possible injuries that an occupant might suffer, regulations related to injury prevention and up-to-date technologies that contribute to crash safety improvement and occupant protection. The review work is of great significance in the following three aspects. Firstly, a valuable topic (electric shock prevention by using the winding-based discharge methods) is discovered, laying the foundations for the subsequent studies. Secondly, the ordinary readers can gain a more comprehensive understanding of the EV crash safety-related problems. Thirdly, the review work can provide general guidelines (about crash safety-related issues) for the professionals in their future studies.

Chapter 3
New Winding-Based Discharge Strategy for EV Powertrains with Extreme Parameters

This chapter aims to tackle two challenges when using the winding-based discharge techniques to improve the crash safety of EVs, which include first, the mechanism of the winding-based (current injection-based) discharge method is not clear, and second, due to system parameters, the traditional LDA-CI discharge strategy might not be applicable for some high-voltage PMSM-based EVs when the required discharge period is 5 s. In this chapter, firstly, a physical energy flow model (EFM) is proposed to explain explicitly the mechanism of the winding-based discharge methods, making it well-founded to understand and develop novel winding-based discharge strategies. Then, a new winding-based method relying on piecewise d, q-axis current loci is proposed for the PMSM drives with slightly large inertia and relatively small system safe current to quickly reduce the DC-bus voltage during crash. Finally, the proposed discharge algorithm is verified by experiments.

3.1 Introduction

As for the traditional LDA-CI discharge strategy in [89] and [95–97], of which equivalent circuit is shown in Fig. 3.1, the features of its implementation procedures include that firstly, when a severe crash occurs, a fixed large non-zero d-axis current (negative value) is injected into the machine, and secondly, the q-axis current is controlled to maintain at zero for the sake of generating no braking torque [97]. The expected discharge process for the traditional LDA-CI discharge method is shown in Fig. 3.1 (red curve). It can be seen that immediately when the current is injected into the machine, the DC-bus voltage is expected to experience a sudden drop under 60 V due to the flux-weakening effect, though the machine speed does not change dramatically. Definitely, this process satisfies the best (five-second) discharge requirement, which is introduced in Sect. 2.2.2.

C. Gong, *Crash Safety of High-Voltage Powertrain Based Electric Vehicles*, Springer Theses, https://doi.org/10.1007/978-3-030-88979-1_3

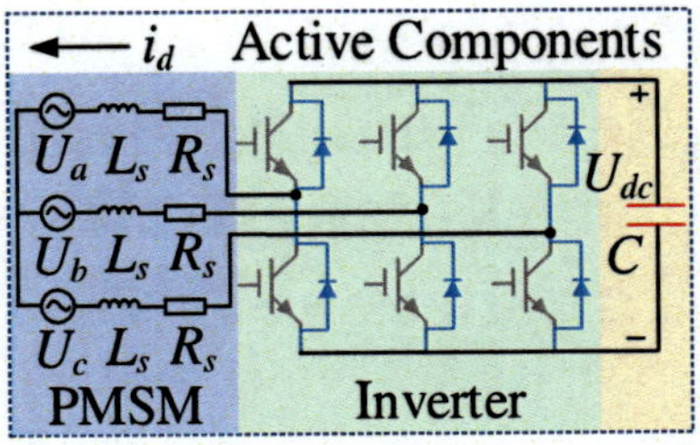

Fig. 3.1 Equivalent circuit and mechanism of LDA-CI discharge strategy

However, when using the traditional LDA-CI method, the DC-bus capacitor discharge characteristics are closely related to the system parameters, such as the system safe current I_{safe} which neither causes magnet demagnetization nor exceeds the maximum allowable current (determined by power electronic devices and system thermal properties), and rotor inertia J. Specifically, when J is large but I_{safe} is relatively small (e.g., $J = 0.24$ kg m^2 and $I_{safe} = 100$ A), the expected discharge performance cannot be achieved. In detail, the discharge time (the time before bus voltage reduces to 60 V) will exceed 5 s even when the injected current is set as the lower limit of the safe level ($-I_{safe}$), which does not satisfy the best discharge requirement (blue curve in Fig. 3.2). Considering that the occupants are more likely to suffer electric shock injury as the discharge period increases, it is significant to develop new winding-based discharge strategy to achieve five-second discharge for those systems with large rotor inertia and relatively small system safe current.

One potential effective way to shorten the discharge time is to apply a large d-axis current as well as a non-zero negative q-axis current to the motor during discharge, by which the motor speed can drop quickly. This strategy can be denoted as a non-zero d-axis, non-zero q-axis (NDNQ) method. However, a large negative q-axis current will increase the rate of energy conversion from the mechanical energy to the electric one that flows back to the DC-bus side. In this case, the motor windings might not be able to expend the energy synchronously, and a huge voltage surge will emerge inevitably

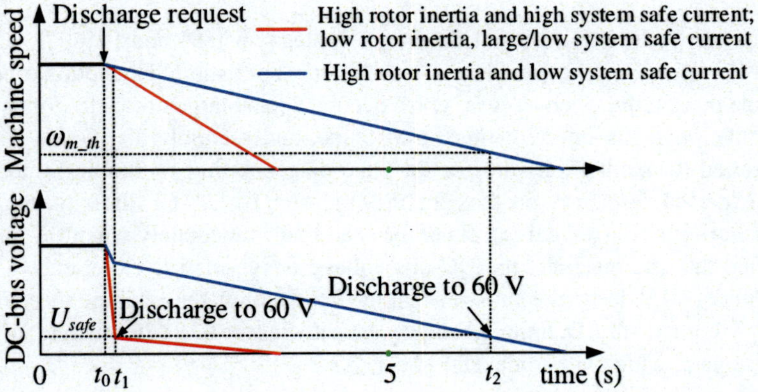

Fig. 3.2 Discharge characteristics for different powertrains

because the capacitance of the DC-bus capacitor is usually not large [117]. In order to avoid excessive voltage surge, it is of great importance to figure out how to control the q-axis current. Before designing a new winding-based discharge strategy, it is essential to understand its working mechanism. Nevertheless, there are no relevant studies that have clearly explained it, and this should be treated as another crucial challenge.

In the rest of this chapter, firstly, an energy flow model (EFM) which is a combination of energetics and electromagnetics is proposed. It deserves to be mentioned that this model is well suited for capturing the transient discharge behaviors of a PMSM drive system, and relying on it, the mechanism and characteristics of both the traditional and novel winding-based discharge schemes are discussed. Secondly, a new winding-based discharge method is proposed to quickly discharge the DC-bus capacitor and safely reduce the motor speed without voltage surge for the high-voltage PMSM powertrains with slightly large inertia and relatively small system safe current. This method adopts piecewise d, q-axis current loci, and it makes full use of the maximum discharge capacity of the system.

3.2 EFM and Mechanism of Winding-Based Discharge Methods

3.2.1 *Energy Flow Model*

Based on the EV powertrain in Fig. 2.3, the EFM used for discharge is established as in Fig. 3.3. The capacitor is be made analogy with a triangular energy tank, and the area of the shaded triangle represents the electric energy packed into the tank. The height of the shaded triangle is equal to the capacitor voltage. v_1 and v_2 represent two valves whose states can be controlled by d, q-axis currents. f_1 is the energy consumed by mechanical friction. f_2 is the kinetic energy of the rotor that is going to be converted into electric energy flowing back to the tank, and most of the energy in the tank is expended in the form of winding heat loss (f_3).

Assume that the discharge is requested at t_0 when the capacitor voltage is U_{dc0}. The initial energy Q_0 stored in the tank is as follows:

$$Q_0 = \frac{1}{2} C \cdot U_{dc0}^2 \tag{3.1}$$

After a period of Δt, the energy flow can be expressed as:

$$f_1 = \int_{\Delta t} F \cdot \omega_m^2 dt \tag{3.2}$$

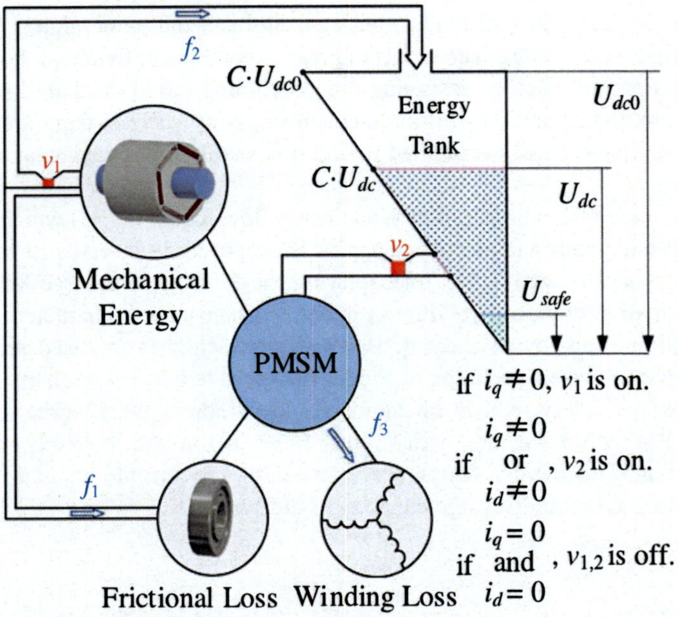

Fig. 3.3 Energy flow model for winding-based discharge strategies

$$f_2 = \frac{1}{2} J \cdot (\omega_{m\_iq\_st}^2 - \omega_{m\_iq\_ed}^2) - f_1 \tag{3.3}$$

$$f_3 = \int_{\Delta t} i_d^2 \cdot R_s dt + \int_{\Delta t} i_q^2 \cdot R_s dt \tag{3.4}$$

It deserves to be addressed that $\omega_{m\_iq\_st}$ and $\omega_{m\_iq\_ed}$ are the rotor mechanical angular speeds before and after i_q is injected to the machine, respectively. This means only when i_q is non-zero (negative) could f_2 exist during discharge. Then, the remaining energy Q_{rem} in the capacitor is:

$$Q_{rem} = Q_0 + f_2 - f_3 \tag{3.5}$$

During discharge, the real-time DC-bus capacitor voltage U_{dc} is:

$$U_{dc} = \sqrt{\frac{2 \cdot Q_{rem}}{C}} \tag{3.6}$$

The above equations are the energetic parts of the proposed EFM, which is based on the law of conservation of energy. However, that is not enough for accurately evaluating the transient performance characteristics of the drive system and calculating the amount of energy conversion. To complement the deficiency, electromagnetic parts are added to the EFM, which are the PMSM model in the direct-quadrature (*d*, *q*) rotating reference frame [118–121].

$$\frac{di_d}{dt} = -\frac{R_s}{L_d}i_d + \frac{L_q}{L_d}p\omega_m i_q + \frac{u_d}{L_d} \tag{3.7}$$

$$\frac{di_q}{dt} = -\frac{L_d}{L_q}p\omega_m i_d - \frac{R_s}{L_q}i_q + \frac{u_q}{L_q} - \frac{\Psi_f}{L_q}p\omega_m \tag{3.8}$$

$$\frac{d\omega_m}{dt} = \frac{1}{J}(1.5p(\Psi_f i_q + (L_d - L_q)i_d i_q) - F\omega_m) \tag{3.9}$$

3.2.2 *Mechanism of Winding-Based Discharge Methods*

Relying on the EFM, the winding-based discharge mechanism (from the perspective of energy flow and consumption) is illustrated in Table 3.1, which can be summarized as:

(a) When both i_q and i_d are zero, v_1 and v_2 are off. There is only one path to consume the rotor kinetic energy by f_1, while the electric energy in the tank cannot flow out. Namely, $f_2 = f_3 = 0$. The capacitor voltage remains at the maximum level U_{dc0}. In this case, the discharge time will be very long and the PMSM will freely speed down.

(b) When i_q is kept at zero and i_d is negative, v_1 is off but v_2 is on. The rotor mechanical energy cannot be converted into electric energy, and it is consumed only by friction, but the energy stored in the tank can be rapidly dissipated because of the existence of winding loss ($f_2 = 0, f_3 \neq 0$). In this case, another typical feature of the system is that there is no energy (regardless of external and internal energy) flowing to the DC-bus capacitor but the PMSM is able to provide voltage support for the DC-bus side. Hence, the DC-bus capacitor voltage will equal the terminal voltage and approximately follow the performance characteristics of the back EMF of the PMSM [122], that is,

Table 3.1 Mechanism of winding-based discharge strategies

Current states		Valve status		Proportion of energy flow		
i_d	i_q	v_1	v_2	f_1	f_2	f_3
Zero	Zero	Off	Off	Large	Zero	Zero
Nonzero	Zero	Off	On	Large	Zero	Nonzero
Zero	Nonzero	On	On	Small	Nonzero	Nonzero
Nonzero	Nonzero	On	On	Small	Nonzero	Nonzero

$$U_{dc} = \sqrt{u_d^2 + u_q^2} \approx \sqrt{3}C_e'(\Psi_f + L_d i_d)\omega_m \tag{3.10}$$

(c) When i_q is nonzero and i_d is zero, both v_1 and v_2 are on. Most of the rotor mechanical energy will be transformed into capacitor electric energy, and then consumed by the windings together with the initial energy in the tank ($f_2 \neq 0$, $f_3 \neq 0$). At the same time, a small percentage of kinetic energy is consumed by the path of f_1. When negative i_q and i_d are injected into the machine, v_1 and v_2 are in open states. A large amount of rotor mechanical energy floods into the tank ($f_2 \neq 0, f_3 \neq 0$) but the capacitor energy can be dissipated by f_3 very quickly.

It deserves to be mentioned the implementation of a winding-based discharge method always contains one or more of the aforementioned cases. For example, as for the traditional LDA-CI discharge method, because the *d*-axis current reference is set as a large negative constant, and the *q*-axis current reference is zero from the very beginning to the end, only case (b) is adopted to achieve the bus capacitor discharge for this method. According to the mechanism shown in Table 3.1, *d*, *q*-axis currents play a critical role in achieving a winding-based discharge strategy because it determines the energy flow process. Hence, when designing a new winding-based discharge method, one of the most important tasks is to properly control the currents to be injected into the machine. The proposed method in this chapter will utilize case (d) for discharge, which will be detailed in Sect. 3.3.

3.3 Winding-Based Discharge Strategies for Systems with Extreme Parameters

For the purpose of redesigning an effective and high-efficiency winding-based discharge approach, it is meaningful to discover beforehand the characteristics and defects of the traditional winding-based bleeding schemes. To achieve this purpose, a case study is conducted in this section. By the use of EFM and Matlab/Simulink, the characteristics of the traditional LDA-CI method are analyzed. In addition, to intuitively verify that a NDNQ method is able to quicken the discharge process or has stronger discharge capacity, an easy-to-get winding-based discharge method by using fixed *d*, *q*-axis currents (denoted as classic NDNQ method) is discussed comparatively. Further, to solve the problems existing in the traditional LDA-CI and classic NDNQ methods, an effective NDNQ method that employs piecewise *d*, *q*-axis currents is designed, of which characteristics are analyzed as well.

A PMSM powertrain with parameters in Table 3.2 is used for case study. In this research, it is assumed that a discharge request occurs when the motor speed is at the rated value ω_{rated}, which is regarded as the maximum speed.

Table 3.2 Parameters of PMSM powertrain

Parameter	Value	Unit
Stator winding resistance R_s	0.275	Ω
d, q-axis inductance L_d, L_d	0.8	mH
The number of pole pairs p	3	–
Moment of inertia J	0.24	kg m^2
Viscous coefficient F	0.0035	–
Permanent magnet flux linkage Ψ_f	0.18	Wb
DC-bus voltage U_{dc0}	310	V
System safe current I_{safe}	100	A
Rated speed ω_{rated}	345	rad/s
DC-bus capacitor C	560	μF

3.3.1 Analysis of Traditional LDA-CI and Classic NDNQ Methods

When the traditional LDA-CI method is applied to the system with slightly large inertia and relatively small safe current, the initial DC-bus voltage is promptly reduced. But it cannot drop under the safe level as quickly as expected in Fig. 3.1, so the capacitor can be charged when the back EMF is higher than the DC-bus capacitor voltage, slowing the discharge rate. The simulation results of capacitor voltage and speed characteristics in Fig. 3.4a prove the above statement. The discharge is requested at 0.5 s when −100 A d-axis and zero q-axis currents are utilized

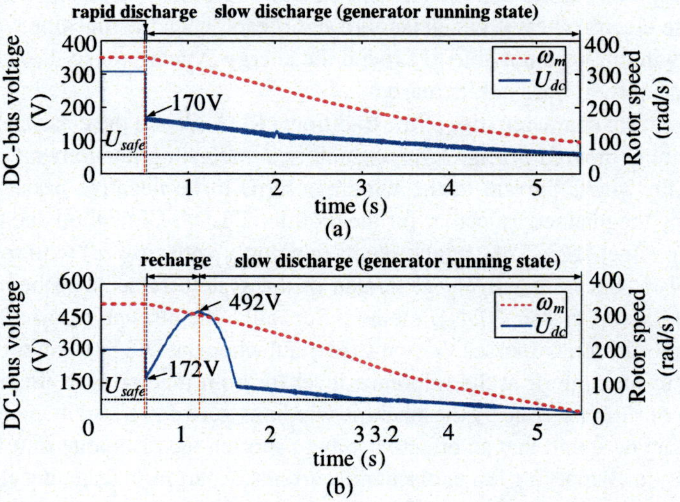

Fig. 3.4 Capacitor voltage and speed characteristics of the winding-based discharge algorithms. **a** Traditional LDA-CI method. **b** Classic NDNQ method

(complying with Fig. 3.1). The capacitor voltage immediately jumped down to 170 V, after which it declines slowly and reaches the safe value until 5.5 s (the discharge time is 5.0 s) when the speed reaches 98 rad/s. The reason why the capacitor voltage can only decrease to 170 V in the rapid discharge process, instead of decreasing to under 60 V as desired is that the highest allowable current of this system is so low that the instantaneous discharge capacity is limited. In detail, when the negative i_d (−100 A) is applied, the air-gap field of the motor is weakened instantly, reducing the back EMF simultaneously to about 170 V at the rated speed. At the moment, the capacitor voltage is much higher than the back EMF, so that the energy can flow in one direction from the capacitor to PMSM. But soon the capacitor voltage will not be constantly higher than the back EMF due to the gradually declining rotor speed. Afterwards, the capacitor will be in a state of discharge accompanied by charge (slow discharge), and only when the rotor speed goes down to the position where the back EMF is equal to U_{safe} will the capacitor voltage remain under 60 V.

The above-mentioned voltage and speed characteristics of the traditional LDA-CI method indicate that it is impracticable to apply a traditional LDA-CI algorithm to the system with slightly large inertia and relatively small safe current when "five-second discharge" is requested. To shorten the discharge time, the *q*-axis current is tried to be controlled to maintain at a constant higher level (classic NDNQ method) so as to speed up the deceleration process. Figure 3.4b illustrates the simulation results of the process. In the situation, a large *d*-axis (−98 A) and a non-zero *q*-axis current (−20 A) are adopted as the reference values at 0.5 s. Compared to Fig. 3.4a, the DC-bus voltage reaches 60 V more quickly within 3.2 s (the discharge time is 2.7 s) when the speed arrives at about 140 rad/s. However, there is an obvious voltage surge in a short period after the beginning of discharge, and if the reference value of i_q is increased, the peak of the voltage surge will get higher. This happens because a larger negative *q*-axis current can increase the energy conversion rate from the mechanical energy to electric energy, which is more significant when the machine speed is high, but the windings are not able to expend the energy synchronously because the total discharge current does not increase.

The current characteristics of the traditional LDA-CI and the classic NDNQ algorithms are compared in Fig. 3.5, in which two interesting phenomena can be seen. Firstly, the *q*-axis current in the windings turns to be negative passively without following the planned trajectory for the traditional LDA-CI method due to the cross-coupling effect [123, 124], which seems to comply with case (d) with regards to the mechanism, but i_q is actively generated in the system rather than being passively injected like the classic NDNQ method. Secondly, the discharge current, especially the *d*-axis current, witnesses a downward trend when the DC-bus voltage is not large enough to maintain it at the reference level for both methods. At the moment, the energy consumption rate by the machine windings goes down simultaneously. Therefore, it can be found that an effective way to shorten the discharge time is to extend the duration of applying larger discharge currents. Apart from faster deceleration, it is also in that way for the classic NDNQ method to achieve a shorter discharge period. In detail, firstly, even though i_d falls just by about 2 A (2%), it lasts 0.8 s longer at the desired position before declining. Secondly, i_q can remain at nearly −20 A during the

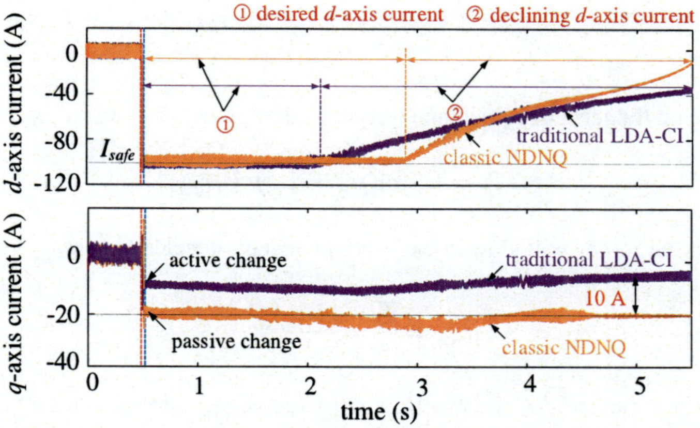

Fig. 3.5 d, q-axis current characteristics of the traditional winding-based algorithms

whole process, and only when the bus voltage reduces almost to zero will it decline for the classic NDNQ method. Comparatively, the q-axis current generated in traditional LDA-CI method is lower (by nearly 10 A) and experiences a declining trend earlier from about 2.5 s. In terms of the defects, the discharge time for the traditional LDA-CI method is not qualified when "five-second discharge" is requested. Although the classic NDNQ method shows more attractive characteristics in this aspect, namely, shorter discharge time, the voltage recovery phenomenon is unwanted for the sake of system safety. Consequently, Sect. 3.3.2 will design a novel winding-based DC-bus capacitor discharge strategy by calculating and applying appropriate q-axis current references to shorten the discharge period, and simultaneously avoid voltage surge.

3.3.2 *Proposed Winding-Based Discharge Method*

By comparing the discharge properties of the above two algorithms, it is feasible to shorten the overall discharge time by using a relatively high negative q-axis current reference i_{q_ref}. However, it is also found that when the q-axis current in the machine is about -10 A from the beginning, there is no voltage rise phenomenon. But when it is controlled at -20 A, the voltage surge shows up in the high-speed range. Hence, the locus of i_{q_ref} should be well designed if a high q-axis current control strategy is employed to shorten the discharge time as well as eliminate the voltage surge. Now, a new issue of how to determine the reference signal arises.

When the negative reference i_{q_ref} and d-axis current reference i_{d_ref} keep fixed, a constant electromagnetic braking torque T_e will be produced in the PMSM. Assuming that the real currents in the machine can be controlled to track the reference values, it can be derived that:

$$T_e = 1.5p(\Psi_f i_q + (L_d - L_q)i_d i_q) = 1.5p(\Psi_f i_{q\_ref} + (L_d - L_q)i_{d\_ref} i_{q\_ref}) \quad (3.11)$$

Ignoring the term of reluctance torque for simplicity, T_e can be expressed as:

$$T_e = 1.5p\Psi_f i_q = 1.5p\Psi_f i_{q\_ref} \quad (3.12)$$

Then, the motor will slow down with a constant deceleration a_{dec}.

$$a_{dec} = \frac{T_e}{J} = \frac{1.5}{J} p\Psi_f i_{q\_ref} \quad (3.13)$$

After a period of Δt, the kinetic energy converted into the electric energy f_2 in (3.3) can be rewritten as:

$$\begin{aligned} f_2 &= \frac{J[\omega_{m\_iq\_st}^2 - (\omega_{m\_iq\_st} + \frac{1.5p\Psi_f i_{q\_ref}}{J} \cdot \Delta t)^2]}{2} - f_1 \\ &= -\frac{3}{2} p\Psi_f i_{q\_ref} \omega_{m\_iq\_st} \Delta t - \frac{9}{8J}(p\Psi_f i_{q\_ref})^2 \Delta t^2 - f_1 \end{aligned} \quad (3.14)$$

It can be noted that f_2 is proportional to the machine speed if a constant q-axis current is adopted. Thus, it can be deduced that the phenomenon that the energy conversion rate (from mechanical energy to electric energy) is higher than the energy consumption rate by using machine windings ($f_2 > f_3$) is more inclined to happen over the high-speed range. The problem can be solved by employing piecewise q-axis current as in Fig. 3.6. The entire discharge process is divided into several periods of Δt. i_{q_ref} is smaller when the speed is high, and it increases incrementally as the machine speed gets down. On the basis of EFM, the key technique proposed

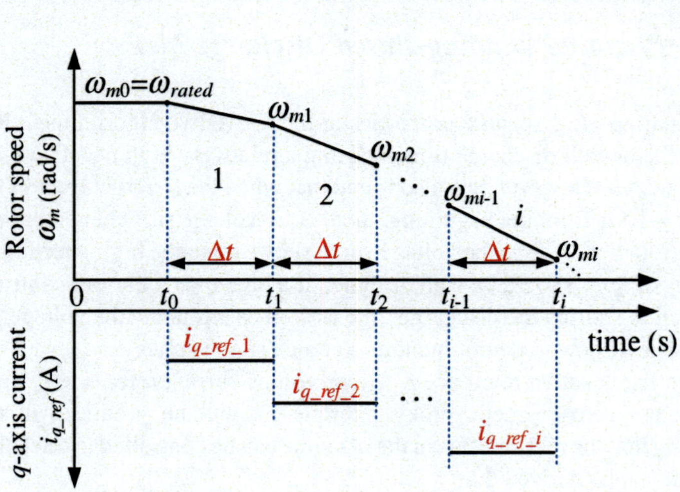

Fig. 3.6 q-axis current reference and desired speed characteristics in the entire discharge process

in this study to avoid voltage surge can be described as: letting $f_2 \leq f_3$ within each period of Δt. In other words, that phenomenon will disappear as long as the mechanical energy conversion rate is not higher than the energy consumption rate. The procedures for computing $i_{q\_ref\_i}$ for each segment are given below, where i represents the number of segment and $i = 1, 2, 3, \ldots$ It needs to be mentioned that because most of the residual energy is consumed by the machine windings for a winding-based discharge strategy, the energy consumed by mechanical friction (f_1) is assumed to be ignored subsequently for the sake of simplicity. Another reason why this assumption is reasonable is that if a winding-based discharge strategy is proven to be effective theoretically without considering friction, the actual discharge process must satisfy the requirement (five-second discharge) because friction could increase energy consumption rate (explained at the end of this section).

In this study, the author tries to take advantage of the maximum discharge capacity of the system, so the system current is expected to be controlled to maintain at I_{safe} during the whole process, that is,

$$i_{d\_ref\_i}^2 + i_{q\_ref\_i}^2 = I_{safe}^2 \tag{3.15}$$

Taking the ith interval as an example, on the basis of the formulas (3.4) and (3.15), it can be derived that:

$$f_3 = I_{safe}^2 \cdot R_s \cdot \Delta t \tag{3.16}$$

Besides, within the ith interval (see Fig. 3.6), substitute the initial speed ω_{mi-1} and q-axis current $i_{q\_ref\_i}$ into (3.14), it can be obtained that:

$$f_2 = -\frac{3}{2} p \Psi_f i_{q\_ref\_i} \omega_{mi-1} \Delta t - \frac{9}{8J} (p \Psi_f i_{q\_ref\_i})^2 \Delta t^2 \tag{3.17}$$

According to the requirement $f_2 \leq f_3$, $i_{q\_ref\_i}$ can be calculated as:

$$\frac{-\omega_{mi-1} + \sqrt{\omega_{mi-1}^2 - \frac{2\Delta t}{J} I_{safe}^2 R_s}}{\frac{1.5 p \Psi_f \Delta t}{J}} \leq i_{q\_ref\_i} < 0 \tag{3.18}$$

As long as the q-axis current satisfies (3.18), the requirement “no voltage surge” can be met. However, in order to decrease the PMSM speed as fast as possible, the magnitude of $i_{q\_ref\_i}$ is supposed to be maximum within the permissible range. Therefore, the required injected q-axis currents are as follows for each period Δt.

$$i_{q\_ref\_i} = \frac{-\omega_{mi-1} + \sqrt{\omega_{mi-1}^2 - \frac{2}{J} I_{safe}^2 R_s \Delta t}}{\frac{1.5 p \Psi_f \Delta t}{J}} \tag{3.19}$$

Substitute (3.19) into (3.15), a piecewise *d*-axis current reference locus can be derived as:

$$i_{d\_ref\_i} = -\sqrt{I_{safe}^2 - i_{q\_ref\_i}^2} \tag{3.20}$$

As for the proposed method, three important aspects need to be addressed. Firstly, because piecewise *d*-axis and *q*-axis currents are used for energy dissipation, it can be called segmented/piecewise NDNQ discharge strategy. Secondly, by carefully looking at Eq. (3.18), the calculated *q*-axis current reference will be imaginary when $\omega_{i-1}^2 < 2I_{safe}^2 R_s \Delta t / J$. This happens only when the machine works in the low-speed or standstill states. Namely, a terrible crash occurs when the EV runs slowly or parks. What deserves to be mentioned is that there is not much residual energy that needs to be consumed during discharge in these cases, and the traditional LDA-CI method is totally qualified. Therefore, the proposed segmented NDNQ discharge method is specially designed for the high-speed ($\omega_{i-1}^2 > 2I_{safe}^2 R_s \Delta t / J$.) situations. Thirdly, it should be acknowledged that the shorter Δt is, the more accurate the discharge method will become. In this study, assume that the change of the machine parameters can be ignored during discharge process and when Δt is set as 0.5 s, Fig. 3.7 shows the values of i_{d_ref} and i_{q_ref} for the different intervals and the corresponding simulation results. In comparison with the traditional LDA-CI method, firstly, the capacitor voltage falls to 60 V within 3.7 s (discharge time is 3.2 s) when the speed gets to about 110 rad/s for the proposed NDNQ method, which becomes shorter and satisfies the discharge requirement. The reason why the threshold speed for the novel method is higher than that for traditional LDA-CI method (but lower than classic NDNQ

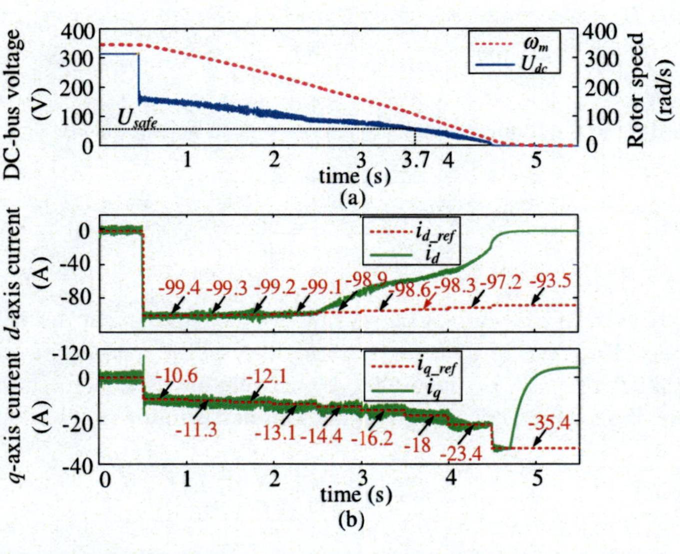

Fig. 3.7 Characteristics of the proposed winding-based algorithm. **a** Capacitor voltage and PMSM mechanical angular speed. **b** *d*, *q*-axis current

method) is that the back EMF of the machine is affected by the flux-weakening d-axis current. The larger the negative i_d is, the higher level the rotor speed should stand at for obtaining the same voltage. Secondly, the q-axis current in the motor can track the reference values well and the d-axis current starts to decline from about 2.5 s, proving that the new current control algorithm is capable of extending the duration of applying relatively large discharge currents to the machine. Then, the voltage surge phenomenon is completely avoided, contributing to eliminating the electrical shock risks of the EV powertrains. Finally, an interesting phenomenon is that the current references are derived by setting $f_2 = f_3$ but the bus voltage continues to drop during discharge. This happens because firstly, the energy consumed by friction (f_1) is ignored when designing the current reference locus. Considering that the mechanical friction can expend energy, the capacitor voltage will decline. Secondly, it can be noticed that at the start of each period in Fig. 3.7b, there exist current overshoots for the real i_q. This will cause that the speed declines faster over the former half period of Δt than that of the latter half period. Consequently, the calculated f_2 is smaller than the real value, which will in turn lead to that the calculated q-axis current reference is relatively lower. Thus, although it is expected that $f_2 = f_3$, f_2 is less than f_3 in reality. But the goal to avoid voltage surge can be achieved. The characteristics of the proposed winding-based method will also be verified by experiments in Sect. 3.4.

3.4 Experimental Results

Experiments are conducted on a system whose parameters are consistent with Table 3.2. It deserves to be mentioned that the original resistance of the PMSM is 0.15 Ω, but it can be regulated to the desired level by directly connecting steel wire with high resistivity for research. The experimental equipment is shown in Fig. 3.8a, of which main hardware was designed by author and provided by collaborators. A DC power supply is available at 310 V. An intelligent power module (IPM), Mitsubishi PM100RLA120, is used as the voltage source inverter with the frequency of 10 kHz. Four thin-film capacitors, DHF DAWNCAP 140 μF, are connected in parallel to compose the desired 560 μF DC-bus capacitor. The proposed discharge algorithm is implemented on DSP TMS320F28335 controller board. The real rotor position is detected by a rotary transformer. Hall current sensors, HNC-100LT, are used to measure three-phase currents while the motor d, q-axis currents are calculated by the digital controller. The DC-bus voltage is measured by a voltage transducer LV25-P. Using the USB-RS485 communication interface, the collected data are transmitted to and further recorded by the host computer. The schematic diagram of the traditional and proposed discharge methods is presented in Fig. 3.8b. Programmatic virtual switches are used to select the operating state of the system. When the system works normally, port 2 is connected and the system can be controlled by any PMSM drive technique, such as double closed-loop speed regulation strategy [125]. Once an emergency occurs, port 1 is connected and the winding-based discharge algorithms are implemented.

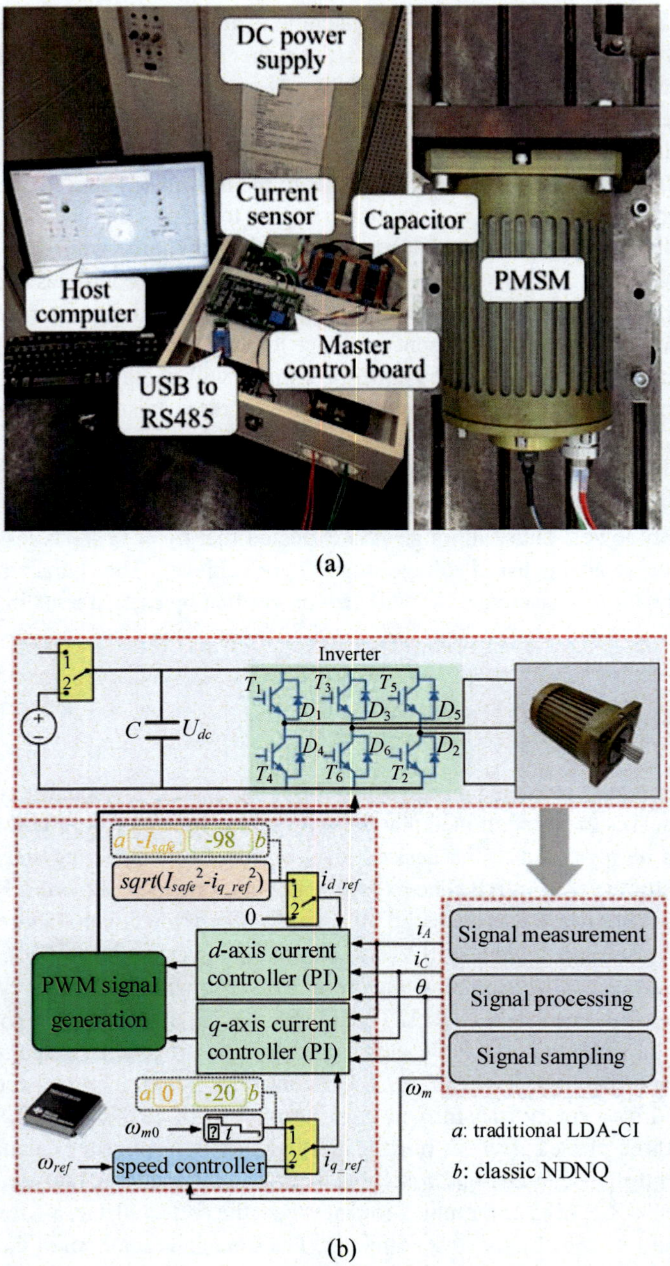

Fig. 3.8 Experimental system. **a** Experimental equipment. **b** Schematic diagram

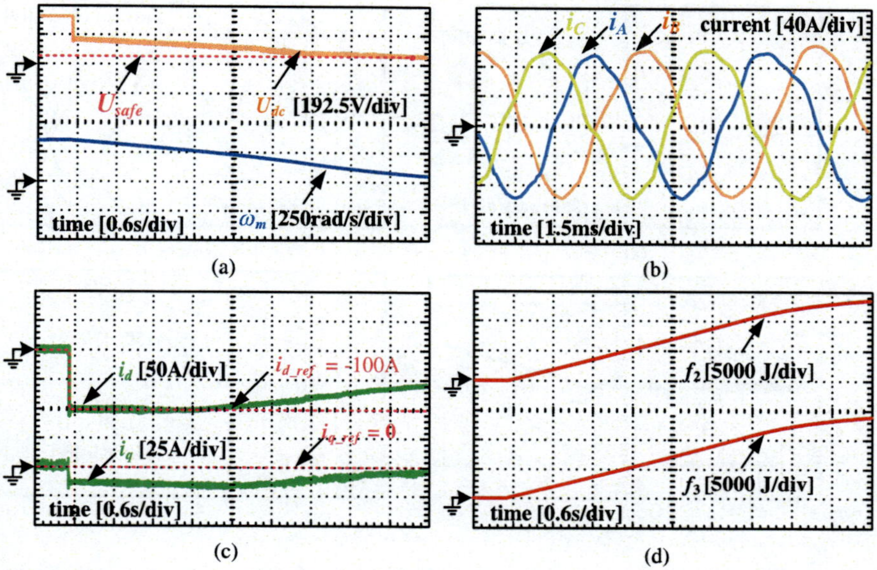

Fig. 3.9 Experimental results of the traditional LDA-CI method when crash occurs at the speed of 345 rad/s. **a** DC-bus capacitor voltage and machine speed. **b** Phase current. **c** *d*, *q*-axis current. **d** Amount of energy conversion and consumption by windings

Assume the discharge request occurs at 0.5 s when the motor speed is ω_{rated}. Figure 3.9a illustrates that the DC-bus capacitor voltage of the traditional LDA-CI method drops to the safe level at nearly 5.75 s, being a little bit longer than that in the simulation result. Figure 3.9b, c show the characteristics of the three phase currents from 0.52 to 0.5215 s and the *d*, *q*-axis currents, respectively. Before 2.4 s, the *d*-axis current can track the prescribed trajectory well, after which it goes down gradually. In accordance with the simulation results, the *q*-axis current (about 14 A) is generated automatically at the time when the discharge is requested, while it starts to decline from 2.5 s and reaches below −5 A at 6.0 s. Additionally, f_2 and f_3 can be calculated according to the EFM, and Fig. 3.9d demonstrates that about 14,000 J of mechanical energy is converted into the electric energy and then consumed by the windings by the end of discharge.

Figure 3.10 depicts the discharge characteristics of the classic NDNQ method. The discharge time is 2.8 s, which is very close to that in the simulation result in Fig. 3.4b. However, a marked voltage surge (about 460 V) appears in the high-speed range. In terms of the current characteristics, i_q can stay at the desired value during the whole process while i_d witnesses a downward trend from 3.1 s. Similar to the traditional LDA-CI method, a total of approximately 14,000 J of energy is expended by the windings during the whole process.

Figure 3.11 demonstrates the experimental results of the proposed winding-based discharge algorithm. Overall, it can be noticed that the DC-bus capacitor voltage drops to 60 V at nearly 3.5 s and the voltage surge disappears. The reason why the

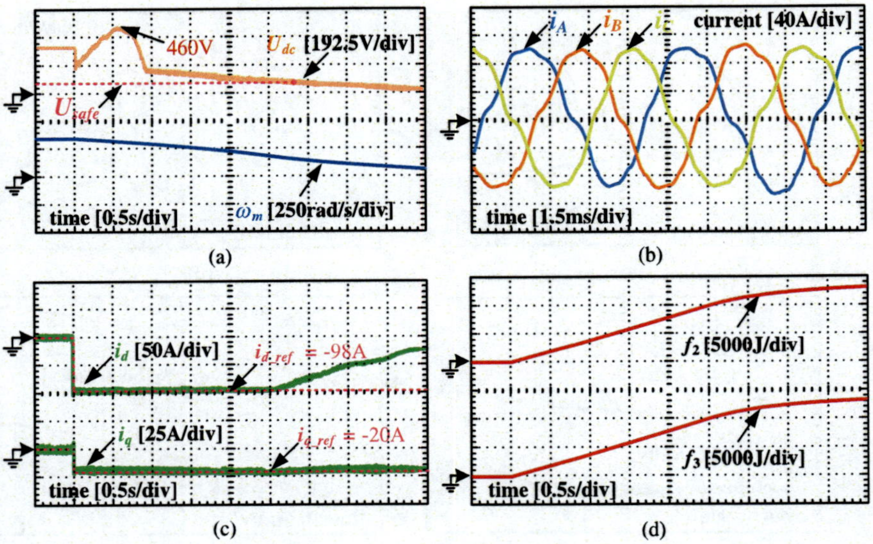

Fig. 3.10 Experimental results of the Classic NDNQ method when crash occurs at the speed of 345 rad/s. **a** DC-bus capacitor voltage and machine speed. **b** Phase current. **c** d, q-axis current. **d** Amount of energy conversion and consumption by windings

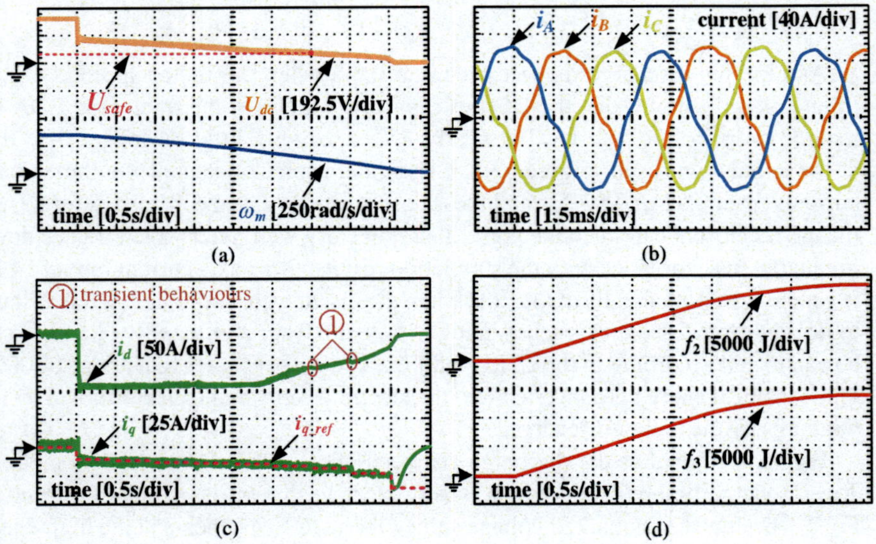

Fig. 3.11 Experimental results of the proposed winding-based discharge algorithm when crash occurs at the speed of 345 rad/s. **a** DC-bus capacitor voltage and machine speed. **b** Phase current. **c** d, q-axis current. **d** Amount of energy conversion and consumption by windings

Table 3.3 Characteristics of DC-bus capacitor discharge algorithms

Type of algorithm	Discharge time (s)	Voltage surge (V)
Traditional LDA-CI method	5.25	None
Classic NDNQ method	2.8	460
Proposed NDNQ method	3.0	None

proposed algorithm can avoid remarkable voltage surge in the PMSM drive system with large inertia is that a piecewise q-axis current locus is used. The higher the mechanical speed is, the relatively lower i_q is applied to the machine, as shown in Fig. 3.11c, producing smaller braking electromagnetic torque. Consequently, the motor speed declines more slowly in the high-speed range, making it possible that the energy consumption rate by windings is no lower than the mechanical energy conversion rate. That stands in contrast with Fig. 3.10c in which i_q always keeps at a high position and the motor windings are incapable of consuming the converted mechanical energy synchronously. Before leaving Fig. 3.11c, it can be found that the d-axis current begins to fall from 2.8 s, proving the statement that the new algorithm can extend the duration of applying relatively large discharge current to the motor compared to the traditional LDA-CI method.

Table 3.3 shows the features of all the three above-mentioned bus voltage bleeding algorithms when they are applied to a drive system with slightly large rotor inertia and relatively small safe current. As far as the proposed method is concerned, although the discharge time is about 3.0 s, which is a little bit longer than 2.8 s for the classic NDNQ method, it has been shortened greatly by 42.8% compared to that in the traditional LDA-CI method. More importantly, the large voltage surge disappears. This is crucial to improve the safety level of both the EV components and passengers during crash.

Moreover, in order to verify that the proposed algorithm is effective regardless of the machine speed, experiment is carried out when assuming a discharge is requested at the speed of 200 rad/s (see Fig. 3.12). For the sake of low initial speed, the calculated q-axis current reference gets lower (−18.5 A) at first. The DC-bus voltage descends to the safe level at about 1.35 s and no large voltage recovery is witnessed. Only about 4500 J of kinetic energy is transferred to electric energy and consumed by the machine windings when the rotor speed arrives at zero.

Looking at the current properties in Figs. 3.11c and 3.12c in detail, the d-axis current fluctuates at the moment when a larger q-axis current is applied to the machine in the proposed algorithm. This phenomenon arises from the transient regulation process of the motor. Besides, the q-axis current can level off at the desired value until the DC-bus voltage is nearly zero, while the d-axis current starts to decline in advance in accord with the simulation results. That indicates that the d-axis current

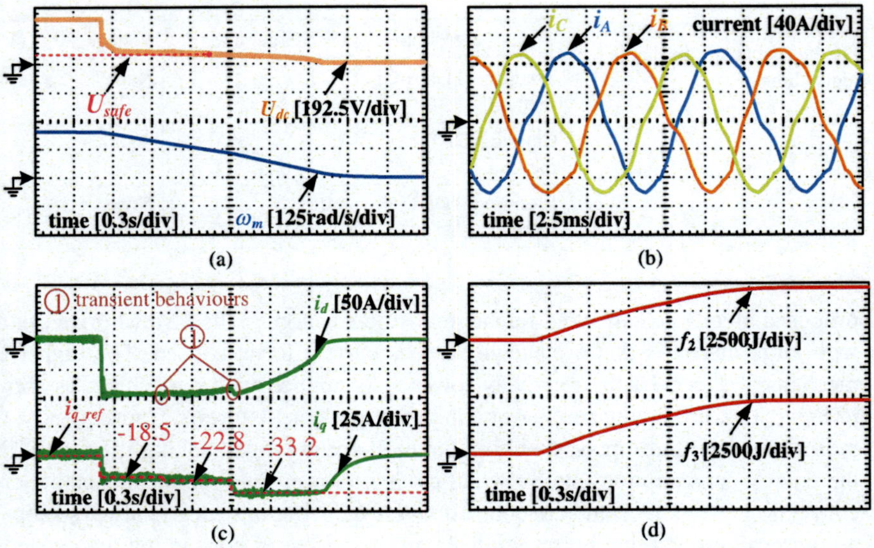

Fig. 3.12 Experimental results of the proposed discharge algorithm when crash occurs at the speed of 200 rad/s. **a** DC-bus capacitor voltage and machine speed. **b** Phase current. **c** *d*, *q*-axis current. **d** Amount of energy conversion and consumption by windings

is more sensitive to the declining DC-bus voltage in the machine when a winding-based discharge algorithm is employed. Therefore, a nonzero *q*-axis current control algorithm is more suitable for discharge the DC-bus capacitor during crash.

3.5 Summary

This chapter proposes an EFM which is a combination of energetics and electromagnetics firstly. The EFM can be used to illustrate the mechanism and characteristics of the winding-based discharge strategies. Secondly, after analyzing the defects of the traditional methods when they are applied to the systems with slightly large rotor inertia and relatively low safe current, a piecewise NDNQ control algorithm is designed to avoid voltage surge and shorten the discharge time simultaneously. The simulation and experimental results prove that, compared to the traditional winding-based discharge methods, the proposed one is more suitable for discharging the DC-bus capacitor of the PMSM powertrains with extreme parameters in EVs during crash because it has stronger discharge capacity. Undoubtedly, the proposed winding-based discharge strategy is able to ensure crash safety of the high-voltage PMSM powertrain-based EVs.

Chapter 4
Hybrid DC-Bus Capacitor Discharge Strategy for EV Powertrains with Highly Extreme Parameters

This chapter aims to solve the challenge that a pure winding-based discharge strategy might not be able to achieve the best discharge process when it is applied to the EV powertrains with highly extreme parameters during crash. Firstly, the defects of the external bleeder-based strategy that is definitely available for discharging the DC-bus capacitor of the powertrains with extreme parameters are discussed analytically, figuring out the necessity of developing a new discharge strategy for those kinds of systems. Then, a hybrid DC-bus capacitor discharge strategy relying on both machine windings and external bleeder circuits is proposed. As for the new strategy, systematic design procedures are developed, which include two main sequential parts: external BR calculation and control algorithm design. Finally, the performance characteristics of the hybrid discharge method are verified by experiment.

4.1 Introduction

Chapter 3 illustrates that the pure winding-based discharge strategies are parameter-dependent. Even though the proposed winding-based discharge strategy in Chap. 3 has stronger discharge capacity than the traditional LDA-CI method, it is just effective when the system parameters are slightly extreme. Once the system parameters become highly extreme, "five-second discharge" cannot be achieved (similar to the blue curve in Fig. 3.1).

To quickly dissipate the residual energy stored in the PMSM powertrains whose parameters are extreme in the crash conditions, as is introduced in Chap. 2, Sect. 2.3.2, an active external bleeder-based discharge method can be employed to achieve the best discharge performance (five-second discharge) as long as the resistance of the BR is designed to be sufficiently small. Theoretically, it is not difficult to design a BR with small resistance. However, when another factor that the power and current withstand levels of the BR should be quite high for the purpose of safety is taken into

C. Gong, *Crash Safety of High-Voltage Powertrain Based Electric Vehicles*, Springer Theses, https://doi.org/10.1007/978-3-030-88979-1_4

account, the size and weight of the bleeding circuit will be incredibly huge, which is unwanted in engineering.

By comparing the winding-based discharge method and the active bleeder-based discharge method, it can be found that firstly, the bleeder-based strategy is not dependent on the system parameters and endowed with enormous dissipation capacity, but it is less competitive considering the size, weight and cost. Secondly, although the winding-based scheme is much more compact and cost-effective, the system parameters have a strong impact on the discharge performance. On these grounds an intuitive hypothesis can be made that these two strategies can be combined together to make up for each other's shortcomings. This chapter proposes a hybrid DC-bus capacitor discharge strategy relying on both machine windings and external bleeder circuits to achieve the best discharge process in the minimum sacrifice of the bleeder size and weight for the EV powertrains with extreme parameters. As for the new method, the machine windings are used as the auxiliary plant for the external bleeder circuits, and it can be applied to any EVs in theory.

4.2 Mechanism and Defects of Bleeder-Based Discharge Method

This part will analyze the mechanism of the bleeder-based discharge method. Then, the current level, resistance, size and weight of the BR are theoretically discussed by means of case study, illustrating that a combined discharge strategy is highly desired. Considering that an EV crash might occur not only in the vehicle running conditions but also in the parking situations, and the bleeder-based discharge strategy should be applicable to them, a comprehensive analysis involving both aspects will be carried out. In order to explicitly explain the working mechanism of the discharge method, equivalent discharge circuits are extracted from Fig. 2.5a for the standstill and running situations, respectively, which are shown in Fig. 4.1.

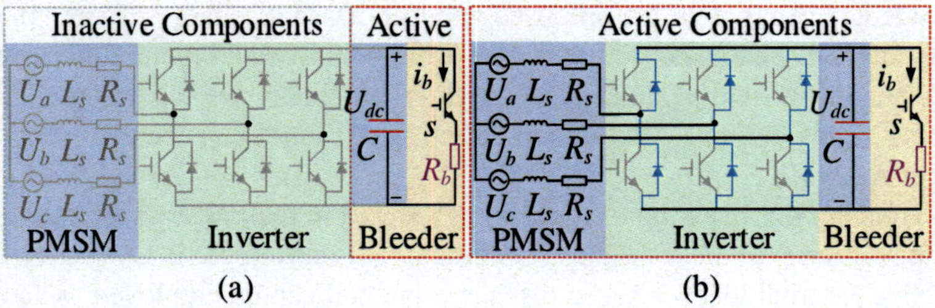

Fig. 4.1 Equivalent circuits and mechanism of bleeder-based discharge method. **a** Standstill cases. **b** Running cases

4.2.1 Mechanism and BR for Standstill Cases

When a discharge request occurs at the machine speed of zero, no residual energy is stored in the motor and only the DC-bus capacitor voltage requires to be diminished. Therefore, the PMSM and inverter are inactive while the capacitor and the bleeder circuits are activated. In this case, the capacitor and the bleeder constitute a simple resistance–capacitance (RC) net, so the real-time capacitor voltage U_{dc} during discharge is represented as:

$$U_{dc} = U_{dc0} \cdot e^{-\frac{t}{R_b C}} \tag{4.1}$$

Taking the required discharge period t_r and the safe DC-bus voltage U_{safe} into account, the required resistance of the BR can be derived as:

$$R_b \leq -\frac{t_r}{C \cdot \ln\left(\frac{U_{safe}}{U_{dc0}}\right)} \tag{4.2}$$

When designing a wound-braking-resistor, apart from its resistance, the current-carrying capacity is another crucial parameter that determines the wire size (diameter and length). During discharge, the total energy Q_b that needs to be dissipated is:

$$Q_b = Q_0 - Q_{safe} = \frac{1}{2}C(U_{dc0}^2 - U_{safe}^2) \tag{4.3}$$

Hence, the root-mean-square (RMS) discharge current i_{bRMS} is:

$$i_{bRMS} = \sqrt{\frac{Q_b}{R_b t_r}} \tag{4.4}$$

4.2.2 Mechanism and BR for Running Case

If an emergency happens when the vehicle runs on the road, the kinetic energy of the PMSM rotor as well as the energy stored in the bus capacitor should be expended by the BR. In the process of energy consumption, the machine works as a generator, inducing a continuously decreasing back EMF that is related the machine speed. As for the inverter, the emergency triggers an inherent protection mode with "shut-down" control signals applied to all of the transistors. Whereas, the six free-wheeling diodes cannot be sealed off, constituting an uncontrolled rectifier (UR). The back EMF of the machine will be rectified by the UR, charging the capacitor and generating the bleeding current i_b that passes through the BR.

In order to reduce the analytical complexity, an appropriate assumption that the effects of the machine inductance and the capacitor capacitance can be ignored since they are small [126]. On this ground the magnitude of the braking q-axis current i_q in the machine equals the bleeding current i_b (active current) and the d-axis current is 0, namely,

$$i_q = -i_b = -\frac{U_{dc}}{R_b} \tag{4.5}$$

Moreover, according to [122], taking the resistance of the machine windings into account, the DC-bus capacitor voltage can be approximated as:

$$U_{dc} = \sqrt{3}C_e^{'}\Psi_f\omega_m - 2i_bR_s \tag{4.6}$$

Then, On the basis of the motor model Eq. (3.12), the braking torque T_e of a PMSM is described as follows:

$$T_e = 1.5p\Psi_f i_q = -1.5p\Psi_f i_b \tag{4.7}$$

And the real-time machine speed can be expressed as:

$$\omega_m = \omega_{m0} + \int_0^t a_{dec}\mathrm{d}t = \omega_{m0} - \int_0^t \frac{T_e}{J}\mathrm{d}t \tag{4.8}$$

Substitute (4.5), (4.7) and (4.8) into (4.6), U_{dc} can be rewritten as:

$$U_{dc} = \frac{\sqrt{3}JR_bC_e^{'}\Psi_f\omega_{m0} - 1.5\sqrt{3}C_e^{'}p\Psi_f^2\int_0^t U_{dc}\mathrm{d}t}{J(R_b + 2R_s)} \tag{4.9}$$

Take the derivative of (4.9), and the voltage descending rate (VDR) can be expressed as:

$$\frac{\mathrm{d}U_{dc}}{\mathrm{d}t} = -\frac{1.5\sqrt{3}C_e^{'}p\Psi_f^2}{J(R_b + 2R_s)}U_{dc}(t) \tag{4.10}$$

Since that U_{dc} experiences a downward trend, according to (4.10), the voltage VDR will decline continuously as well. Therefore, within t_r, it can be derived that:

$$\int_0^{t_r} U_{dc}\mathrm{d}t \geq \left(1 + \frac{0.75\sqrt{3}C_e^{'}p\Psi_f^2 t_r}{J(R_b + 2R_s)}\right)t_r U_{safe} \tag{4.11}$$

Based on (4.11), the resistance of the BR is supposed to meet the following criteria:

$$R_b \leq -\frac{1}{2J(U_{safe} - a_1)}(4JR_sU_{safe} - 2a_1JR_s + a_2U_{safe}t_r - \sqrt{4a_1JR_s(a_1JR_s + a_2U_{safe}t_r) - a_2^2U_{safe}t_r^2(U_{safe} - 2a_1)}) \quad (4.12)$$

where $a_1 = \sqrt{3}C_e'\Psi_f\omega_{m0}$ and $a_2 = 1.5\sqrt{3}C_e'p\Psi_f^2$.

In comparison with the zero-speed cases, in addition to that the bus voltage should decline to U_{safe} within t_r, the machine speed also needs to decrease to the threshold value ω_{m_th} at which the line-to-line back EMF is U_{safe} for the running cases. Hence, the energy to be expended includes not only the electrical part stored in the capacitor but also the kinetic energy stored in the machine rotor.

$$Q_b = \underbrace{\frac{1}{2}J(\omega_{m0}^2 - \omega_{m\_th}^2)}_{Kinetic\ energy} + \underbrace{\frac{1}{2}C(U_{dc0}^2 - U_{safe}^2)}_{Electrical\ energy} \quad (4.13)$$

Then, the RMS discharge current is calculated as:

$$i_{bRMS} = \sqrt{\frac{Q_b}{(R_b + R_s)t_r}} \quad (4.14)$$

4.2.3 *Evaluation of Size and Weight Sacrifice*

In order to implement an intuitive discussion on the properties of the BR, a PMSM powertrain for EV with parameters (original parameters of the system shown in Table 3.2) in Table 4.1 is studied. Currently, one of the most common materials used for making precise braking resistor is the alloy of copper (Cu) and nickel (Ni) [127]. Taking CuNi44 (ISOTAN®) whose resistivity is $\rho_r = 49 \times 10^{-8}$ Ω·m and density is $\rho_m = 8900$ kg/m³ as an example, the minimum wire diameter is selected based on the relationship between the wire diameter d and the current-carrying capacity i_{ca} (see Fig. 4.2). Here, current capacity refers to the safe current that can flow the conductor considering thermal properties.

$$i_{ca} = 0.3516d^2 + 2.6475d - 0.1552 \quad (4.15)$$

Table 4.2 shows the properties of the BR (just the conductor part) when the required discharge time is 5 s. Firstly, it can be seen that the parameters designed for the running mode (initial speed should be set as the maximum value) are suitable for the standstill situations. Then, to satisfy the fastest discharge requirement, although the required resistance is only 7.33 Ω, the diameter d_b of the conductor is 4.5 mm and the length l_b is 237.8 m, weighing around 33.6 kg. However, such an enormous

Table 4.1 Parameters of PMSM Powertrain

Parameter	Value	Unit
Stator winding resistance R_s	0.15	Ω
Number of pole pairs p	3	–
d, q-axis inductance L_d, L_q	0.8	mH
Moment of inertia J	0.24	kg m^2
Permanent magnet flux linkage Ψ_f	0.18	Wb
DC-bus voltage U_{dc0}	310	V
System safe current I_{safe}	100	A
Rated speed ω_{rated}	345	rad/s
Threshold speed ω_{m_th}	65	rad/s
Voltage constant C'_e	2.88	–
Required discharge time t_r	5	s
DC-bus capacitor C	560	μF

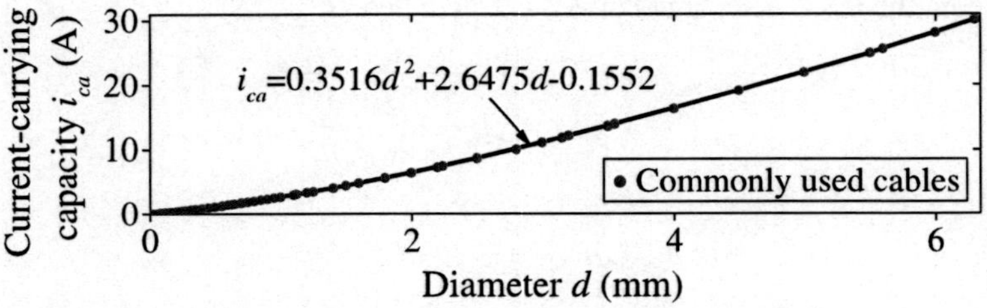

Fig. 4.2 Relationship between conductor diameter and current-carrying capacity

Table 4.2 Parameters of bleeding resistor

Resistor properties	Standstill	Running	Unit
Resistance R_b	≤ 5415	≤ 7.33	Ω
Current level i_{bRMS}	0.03	19.2	A
Minimum diameter d_b	0.022	4.5	mm
Required length l_b	4.2	237.8	m
Calculated mass m_b	0.014·10^{-3}	33.6	kg

resistor is unwelcome because it increases the vehicle weight greatly, so in practice, the lower power-level resistors are usually used, but the discharge time will exceed 5 s.

Overall, the EVs are placing a high demand on the novel discharge strategies characterized by rapid discharge but small size and weight.

4.3 Proposed Hybrid Discharge Technique

Figure 4.3 illustrates the block diagram of the proposed hybrid discharge strategy. Compared to Fig. 4.1, the novel bleeding system is composed of not only the motor, diodes and external bleeder circuits but also the transistors in the inverter. Besides, the vector control algorithms based on specially-designed *d*, *q*-axis current injection are needed to generate the control signals. Similar as Fig. 3.8, programmatic virtual switches are used to select the working modes of the system. This part will give a design method for the BR, and on this ground the bleeder size and weight reductions are analyzed. Then, different discharge regulation modes and algorithms are developed considering the initial condition when a discharge request occurs.

4.3.1 Design of BR for Proposed Discharge Method

When designing the BR for the hybrid discharge system, the extreme condition at which the initial speed ω_{m0} equals the maximum value needs to be considered. It has been stated in Chap. 3 that the maximum machine speed is assumed to equal the rated value ω_{rated} in this study. In this case, the total energy to be dissipated is denoted as Q_{b_des}, which can be calculated by (4.13) when letting ω_{m0} equal ω_{rated}.

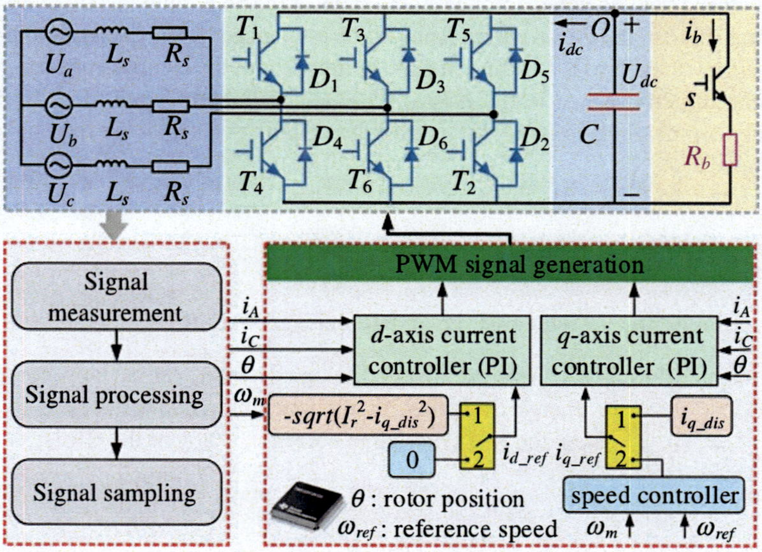

Fig. 4.3 Block diagram of the proposed hybrid discharge method

(a) Resistance of BR

By contrast with the individual bleeder-based discharge method of which braking torque is produced depending on the quotient of the bus voltage and BR, as is shown in (4.7). The deceleration process of the new approach relies on the injected d, q-axis currents (transformed from the phase currents i_a and i_c). Further, another crucial feature of the proposed hybrid discharge technique is that, thanks to the BR, it is not necessary to adopt the piecewise q-axis current reference to avoid the voltage surge phenomenon as long as the resistance is relatively small. The reason is as follows: the voltage surge arises because within a short period, the rotor kinetic energy that is converted into the electric energy (KETEE) is larger than the total bleeding capacity (sum of internal and external dissipation). However, for the hybrid method, once the bus voltage rises greatly, the instantaneous bleeding power of the BR will shoot up and the short-period bleeding energy would surpass the KETEE, preventing the capacitor voltage from continuous growth. Based on these, a constant large q-axis current reference that can ensure the five-second discharge requirement is needed. According to (4.7) and (4.8), the q-axis current for design (i_{qref_des}) within t_r should track the following locus:

$$i_{qref\_des} = \frac{J(\omega_{m\_th} - \omega_{rated})}{1.5pt_r\Psi_f} \tag{4.16}$$

Although the q-axis current in the machine is inclined to remain at the reference level when bleeding enters into the stable state, at the beginning instant of discharge, the real value of the q-axis current can still be influenced by the quotient of the initial capacitor voltage and BR because the DC component of the DC-bus current i_{dc} (active current, directly related to i_q) in Fig. 4.3 is approximately equal to i_b. In order to rapidly compel the q-axis current to trace the targeting value, let the resistance of the BR satisfy:

$$R_b = \frac{U_{dc0}}{|i_{qref\_des}|} \tag{4.17}$$

(b) Conductor design considering BR size and weight

Assume that during discharge, the energy dissipated by the external bleeder is Q_{bex_des} while Q_{bin_des} is the energy consumed by the machine windings, where $Q_{bex\_des} + Q_{bin\_des} = Q_{b\_des}$. Then, the RMS discharge current i_{b_exRMS} of the BR is:

$$i_{b\_exRMS} = \sqrt{\frac{Q_{bex\_des}}{R_b t_r}} \tag{4.18}$$

i_{b_exRMS} is vital to the conductor diameter and length of the BR whose resistance has been decided. Hence, in order to reduce the size and weight of the bleeder circuits as much as possible, i_{b_exRMS} should be minimal. Consequently, when designing the

BR, Q_{bex_des} is supposed to be the lowest in the extreme condition. In this case, the best bleeding capacity of the machine windings needs to be utilized. Based on this, the currents in the machine is expected to be controlled to maintain at the system safe current I_{safe}, which should not pose any potential risks (including thermal damage and overcurrent, etc.) to the system during the whole discharge process, that is,

$$i_{dref\_des}^2 + i_{qref\_des}^2 = I_{safe}^2 \tag{4.19}$$

Then, Q_{bex_des} is calculated by:

$$Q_{bex\_des} = Q_{b\_des} - I_{safe}^2 R_s t_r \tag{4.20}$$

Substitute (4.16)–(4.18) and (4.20) into (4.15), the required diameter of the BR conductor is:

$$d_b = -3.765 + \sqrt{14.615 + 2.843 \sqrt{\frac{J(Q_{b\_des} - I_{safe}^2 R_s t_r)(\omega_{rated} - \omega_{m\_th})}{1.5 p t_r^2 \Psi_f U_{dc0}}})} \tag{4.21}$$

Here, the thermal performance is not further analyzed because i_{ca} takes it into account. And the length is:

$$l_b = \frac{\pi R_b d_b^2}{4\rho_r} \times 10^{-6} \tag{4.22}$$

The mass of the required conductor can be calculated by:

$$m_b = \frac{\pi}{4} \rho_m d_b^2 l_b \cdot 10^{-6} \tag{4.23}$$

(c) Evaluation of size and weight

As for the PMSM drive system whose parameters are in Table 4.1, the properties of the designed BR are shown in Table 4.3. In comparison with Table 4.2, the required BR resistance in the proposed system increases to 18.8 Ω, but the conductor diameter d_b

Table 4.3 Properties of bleeding resistor

Resistor properties	Value	Unit
Resistance R_b	18.8	Ω
Current level i_{b_exRMS}	8.18	A
Minimum diameter d_b	2.4	mm
Required length l_b	173.5	m
Calculated mass m_b	6.98	kg

decreases to 2.4 mm (46.7%) and the length l_b drops to 173.5 m (27%). Importantly, the weight of the BR conductor experiences a significant decline from 33.6 to 6.98 kg.

Although the weight of the proposed method is still high for a practical application, it is obviously smaller than the pure bleeder-based discharge method. Therefore, the advantages of the proposed method can be described as follows: when five-second discharge is required, the BR weight might be still high (at least for the studied system in this research) for the hybrid discharge method, but it must have been reduced compared to the traditional discharge method. In practice, if the weight, size and cost are limited, we can set a longer discharge period (within one minute) as the required discharge time. In this case, the resistance of the BR will be higher but the weight and size can be further reduced.

Overall, the BR in the hybrid discharge structure is much smaller than that in the bleeder-based discharge system.

4.3.2 Discharge Modes and Control Algorithms

After adopting the above-mentioned hybrid DC-bus capacitor discharge topology, the EV powertrain is endowed with the ability to regulate the bus voltage to drop below the safe level within five seconds when the emergency happens at the speed of rated position. However, it is probable that the discharge is requested at the moment when the machine does not reach the highest speed, including the standstill state. Considering the safety of the EV powertrain, especially the fragile transistors in the inverter, it is unnecessary to always control the *d*, *q*-axis discharge currents to remain at the extreme states (i_{dref_des} and i_{qref_des}), and even the winding-based discharge part can be absent. Three types of discharge modes based on ω_{m0} and the corresponding control algorithms are developed in this section (see Fig. 4.4).

(a) **Full-power discharge mode**

When a discharge request occurs at the speed of the maximum value, both the internal machine windings and the external bleeder circuits are needed for dissipating the residual energy. In this case, the hybrid system has to take full advantage of its own discharge capacity and the required discharge current I_r equals I_{safe}. Consequently, the *d*, *q*-axis discharge currents (i_{d_dis} and i_{q_dis}) in the machine should comply with the calculated parameters during BR design:

$$\begin{cases} i_{q\_dis} = i_{qref\_des} \\ i_{d\_dis} = -\sqrt{I_{safe}^2 - i_{q\_dis}} \end{cases} \tag{4.24}$$

It is expected that the DC-bus capacitor voltage will get down to U_{safe} soon by the use of the combined scheme, removing the electrical shock risks. After that, the controller will continue to implement the discharge algorithms until the bus voltage arrives at zero.

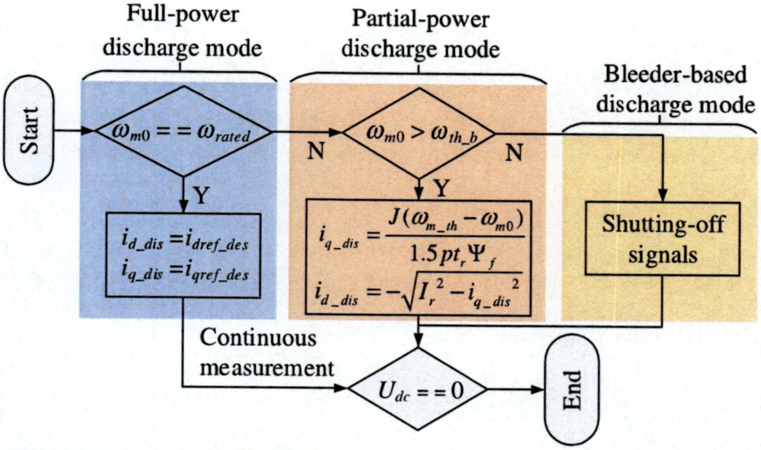

Fig. 4.4 Discharge modes and control algorithms

(b) Partial-power discharge mode

When ω_{m0} is lower than ω_{rated} but higher than the corner threshold ω_{th_b} below which only the external BR is able to pull the bus voltage down to 60 V within 5 s, both the windings and BR have to be adopted for discharge as well, but other than (4.24), the d, q-axis current in the machine can be modest.

Firstly, the fast discharge requirement must be satisfied, so the q-axis discharge current should be controlled as:

$$i_{q\_dis} = \frac{J(\omega_{m\_th} - \omega_{m0})}{1.5p\Psi_f t_r} \tag{4.25}$$

Secondly, as is illustrated in Fig. 4.3, the bleeder circuits together with the bus capacitor constitute the load of the PMSM that works as a generator. Since the capacitance is usually small, the load can be further approximated as a resistor [126, 128]. Then, ignoring the mutual inductance, the current and voltage in a permanent magnet generator can be depicted by an approximated phasor diagram in Fig. 4.5, X_d and X_q are the d and q-axis reactance; $X = X_d + X_q$, representing the machine reactance. $\mathbf{I}_d$ and $\mathbf{I}_q$ are the d and q-axis current vector; $\mathbf{U}$ is the phase voltage, and $\mathbf{I}$ is the current (amplitude is I) in the machine. Then, it can be obtained that:

$$U_{dc} = \sqrt{3}(E\cos\phi - IR_s) = \sqrt{3}\left(\frac{C_e'\Psi_f\omega_m i_{q\_dis} - I^2R_s}{I}\right) \tag{4.26}$$

Between 0 and t_r, the energy dissipated by windings (Q_{bin}) and BR (Q_{bex}) can be represented as:

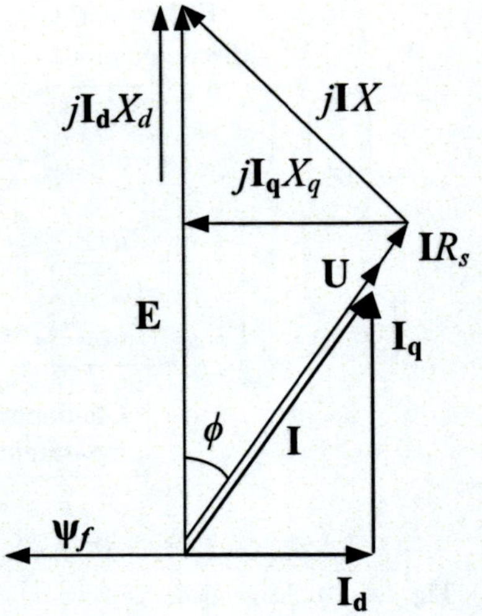

Fig. 4.5 Approximated PMSM phasor diagram operating as a generator

$$\begin{cases} Q_{bin} = \int_0^{t_r} 3I^2 R_s \mathrm{d}t = 3I^2 R_s t_r \\ Q_{bex} = \int_0^{t_r} \frac{U_{dc}^2}{R_b} \mathrm{d}t \end{cases} \tag{4.27}$$

where $Q_{bex} + Q_{bin} = Q_b$. According to (4.7), (4.8) and (4.25)–(4.27), when $\omega_{m0} \leq \omega_{rated}$, the required discharge current I_r in the machine will satisfy the following condition:

$$I_r^2 = \frac{3t_r R_s(ct_r + 2g) + Q_b R_b + \sqrt{\begin{array}{l} Q_b^2 R_b^2 - c^2 t_r^4 R_s(12R_b - 3R_s) + \\ 6R_b R_s t_r[(6gt_r - Q_b)(ct_r - g) + Q_b g] \end{array}}}{3R_s t_r(R_b + R_s)} \tag{4.28}$$

where $c = \frac{1.5pC_e'\Psi_f^2 i_{q\_dis}^2}{J}$ and $g = C_e'\Psi_f\omega_{m0}i_{q\_dis}$. Therefore, the d-axis reference discharge current should be set as:

$$i_{d\_dis} = -\sqrt{I_r^2 - i_{q\_dis}^2} \tag{4.29}$$

Similar as the full-power discharge mode, only when the bus voltage gets down to zero will the implementation stops.

(c) **Bleeder-based discharge mode**

There must be a speed threshold ω_{th_b} below which (including standstill case) there is no need to implement the winding-based discharge algorithms and just the bleeder-based strategy is qualified for the five-second discharge process.

According to (4.5) and (4.6), the bus voltage generated by ω_{m0} is:

$$U_{dc}(0) = \frac{\sqrt{3}C_e' \Psi_f R_b \omega_{m0}}{R_b + 2R_s} \tag{4.30}$$

Obviously, U_{dc} (0) is smaller than U_{dc0} when the initial speed is less than ω_{rated}. But since the resistance of BR is small, the capacitor voltage will quickly decline to the back EMF level, which is called voltage balance phenomenon (VBP, eg., balance time < 20 ms for the aforementioned system). Then, the DC-bus capacitor voltage will be dominated by the induced voltage. On this ground the differential Eq. (4.10) can be solved with the boundary condition U_{dc} (0) after ignoring the balance time.

$$U_{dc}(t) = U_{dc}(0) \cdot \exp\left(-\frac{1.5\sqrt{3}C_e' p \Psi_f^2}{J(R_b + 2R_s)} t\right) \tag{4.31}$$

when $t = t_r$, U_{dc} should be less than U_{safe}. Hence, it can be derived that:

$$\omega_{th\_b} = \frac{U_{safe}(R_b + 2R_s)}{\sqrt{3}C_e' \Psi_f R_b \exp\left(-\frac{1.5\sqrt{3}C_e' p \Psi_f^2}{J(R_b+2R_s)} t_r\right)} \tag{4.32}$$

When $\omega_{m0} < \omega_{th\_b}$, the "shutting-off" signals will be directly applied to the inverter.

Figure 4.6 demonstrates the theoretical DC-bus voltage characteristics under the three different discharge modes. Three main features can be summarized as follows: (1) When the machine windings serve for discharge, apart from ω_{m0}, the negative injected flux-weakening current (*d*-axis current) is another key factor that makes the back EMF of the machine lower than U_{dc0}, resulting in VBP. Therefore, the bus

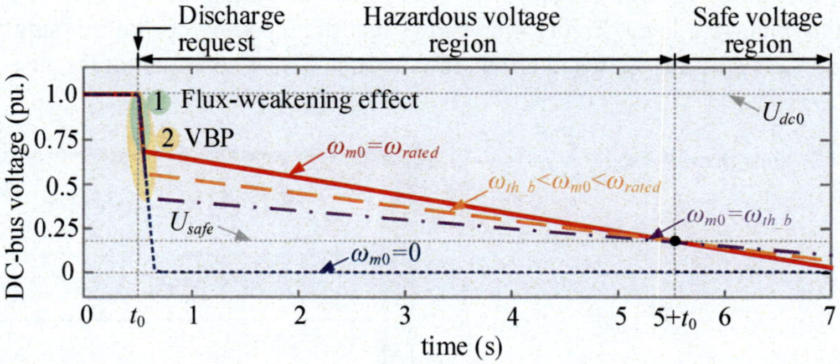

Fig. 4.6 Theoretical DC-bus capacitor discharge characteristics under different control modes

voltage will experience a sharp decrease immediately after the discharge request arises even when the initial speed equals ω_{rated}. (2) When the capacitor voltage is determined by the machine ($\omega_{m0} > 0$), the higher the initial speed is, the larger the VDR becomes at the same discharge moment. (3) In theory, the bus voltage will quickly drop to zero for both the full-power and partial power discharge modes in the safe voltage region. But it lasts long for the bleeder-based discharge mode, which is allowable in this research because the voltage shock risks have been eliminated.

4.4 Experimental Verifications

The experimental equipment is shown in Fig. 3.8. It deserves to be mentioned that the winding resistance shown in Table 4.1 complies with the initial value, so it does not need to be regulated by connecting steel wire with high resistivity for research. In this chapter, the control frequency of the inverter is 7.5 kHz. Moreover, in order to verify the effectiveness of the designed BR resistance, a slide rheochord (see Fig. 4.7) of which resistance can be adjusted to 18.8 Ω (current level 15 A) function as the BR.

Firstly, assume that a discharge request occurs when the motor speed is ω_{rated}. At this moment, the system operates at the full-power discharge mode, and the reference *d* and *q*-axis current reference s can be calculated as − 98.6 A and − 16.5 A, respectively. Overall, Fig. 4.8a illustrates that the capacitor voltage drops to the safe level within around 4.4 s, being slightly shorter than 5 s. This happens because the mechanical friction is ignored when establishing the discharge model. It can be concluded that the proposed BR design and full-power discharge methods are very effective. In terms of the speed characteristics, interestingly, the machine speed is still higher than ω_{m_th} when the DC-bus voltage arrives at 60 V, which is caused by the flux-weakening impact. This also contributes to shortening the discharge time. In accordance with the theoretical analysis, the bus voltage experiences sharp decrease (large VBP) after the discharge algorithms are implemented. In Fig. 4.8b, both the *d* and *q*-axis currents get to the expected level quickly after discharge begins and then, they level off until about 4.9 s when the bus voltage is not able to maintain such high current level. Figure 4.8c shows that the bleeding current passing through the BR (BR current) jumps to about 16 A at first. This phenomenon is caused by the

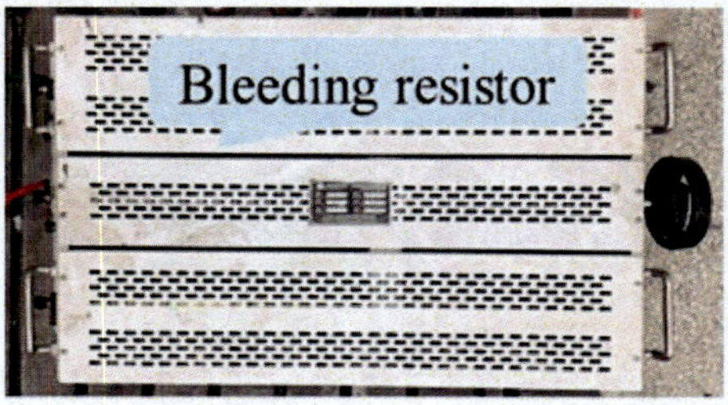

Fig. 4.7 Slide rheostat used as BR

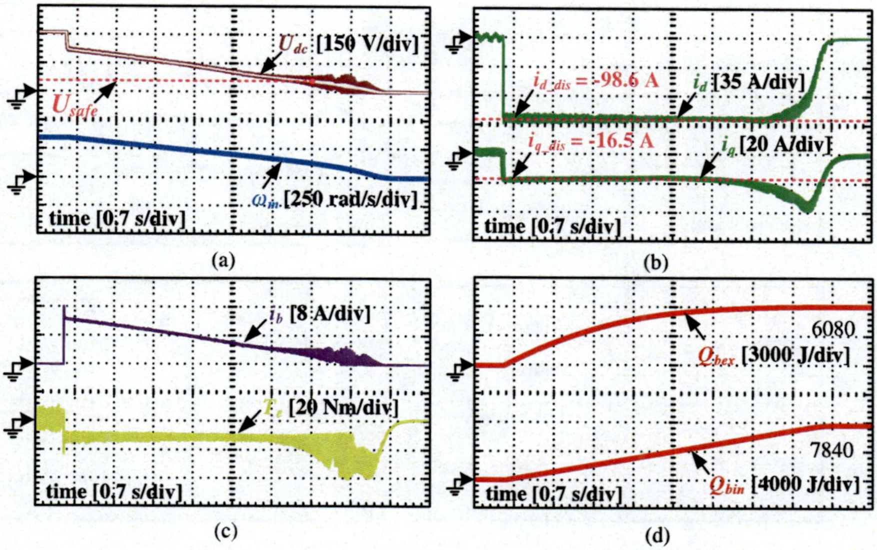

Fig. 4.8 Experimental results of the proposed discharge method when emergency occurs at the speed of 345 rad/s. **a** DC-bus voltage and machine speed. **b** *d*, *q*-axis currents. **c** BR current and braking torque. **d** Energy dissipated by BR and machine windings

initial voltage of the capacitor. Then, the discharge BR current will be dominated by the back EMF, witnessing a linearly downward trend. Moreover, the braking torque is consistent with the *q*-axis current, indicating that the torque calculation method is reasonable. Finally, at the extreme state, a total of 13,920 J (6080 and 7840 J for external BR and internal windings, respectively) are dissipated by the whole system.

Figures 4.9 and 4.10 depict the experimental results when the emergency occurs at the speed of 250 rad/s and 200 rad/s, respectively. In these cases, the partial-discharge mode is employed. In Fig. 4.9, the *d*, *q*-axis discharge currents require to be set as − 54.5 and − 11 A, respectively. The capacitor voltage gets down to the safe level at about 4.8 s (discharge period is 4.3 s). Similar to the extreme state, when the bus voltage is 60 V, the rotating speed is about 98 rad/s which is higher than the threshold. As for the discharge currents, both the *d* and *q*-axis currents can track the reference values well as long as the bus voltage is high enough. The BR current in Fig. 4.9c witnesses a sudden increase at the start of discharge (16 A as well) before it enters into the linear declining region. Figure 4.9d illustrates that about 3705 J energy is consumed by bleeder circuits while 2678 J is dissipated by the machine windings in the form of heat. The VDR in Fig. 4.10a gets smaller than that in Fig. 4.9a because the initial speed has gotten lower, but the discharge period sees little change. When the initial speed is 200 rad/s, the *d*, *q*-axis discharge currents are − 42 and − 8 A, respectively. The braking torque in Fig. 4.10c is smaller than that in Fig. 4.9c, indicating that the discharge process becomes modest.

Figure 4.11 demonstrates the experimental results when the discharge request arises at the speed of ω_{th_b}. Under the bleeder-based discharge method, there are no

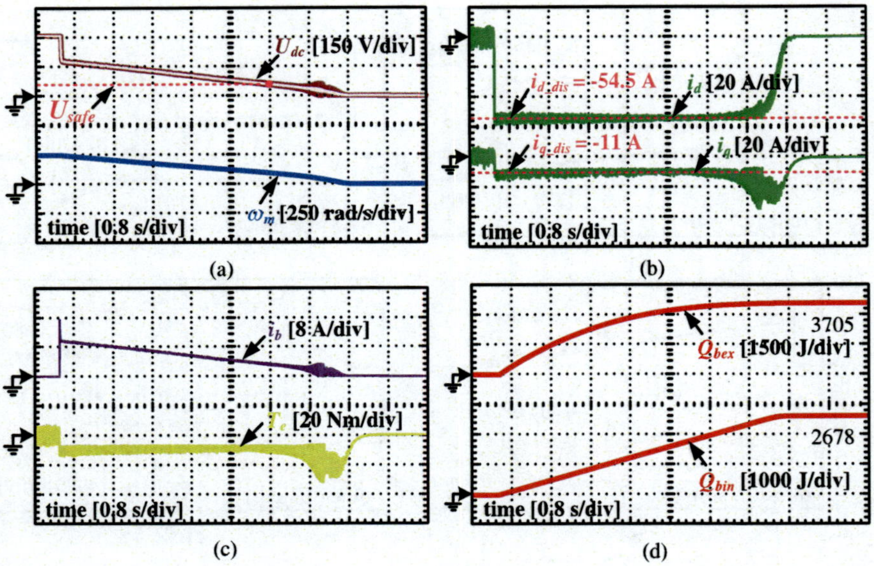

Fig. 4.9 Experimental results of the proposed discharge method when emergency occurs at the speed of 250 rad/s. **a** DC-bus voltage and machine speed. **b** *d, q*-axis currents. **c** BR current and braking torque. **d** Energy dissipated by BR and machine windings

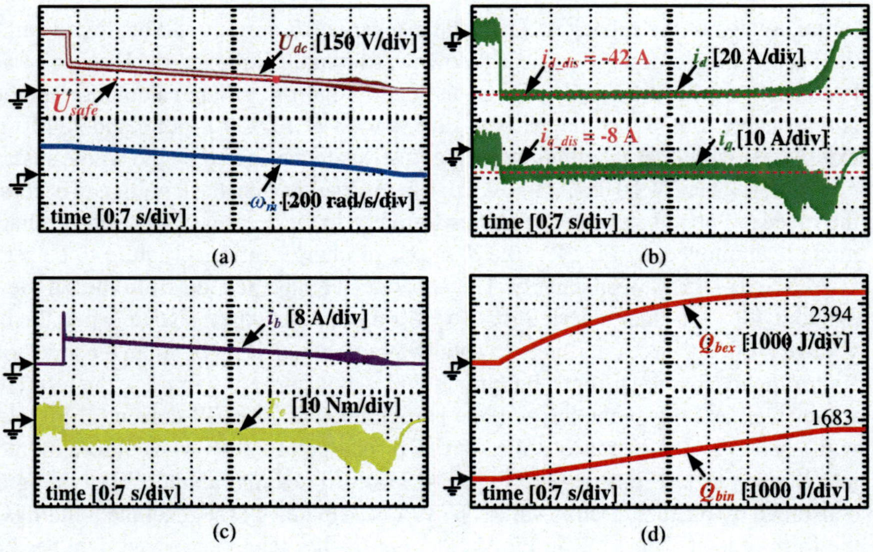

Fig. 4.10 Experimental results of the proposed discharge method when emergency occurs at the speed of 200 rad/s. **a** DC-bus voltage and machine speed. **b** *d, q*-axis currents. **c** BR current and braking torque. **d** Energy dissipated by BR and machine windings

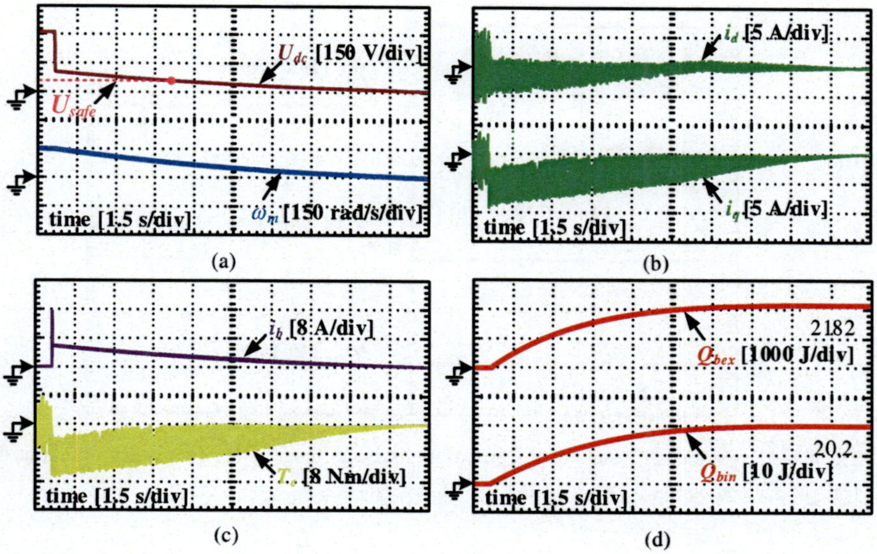

Fig. 4.11 Experimental results of the proposed discharge mode when emergency occurs at the speed of 150 rad/s. **a** DC-bus voltage and machine speed. **b** Discharge *d*, *q*-axis currents. **c** BR current and braking torque. **d** Energy dissipated by BR and machine windings

specially designed currents that are injected into the machine. The induced voltage will generate bleeding current through the UR. Figure 4.11a shows that it is nearly 4.1 s before the bus voltage arrives at the safe level, but the total discharge time is over 10 s. This happens because the braking torque which is related to the bleeding current remains declining (see Fig. 4.11c), leading to the speed deceleration gets lower. Figure 4.11b presents that the *d*-axis component is much smaller than the *q*-axis component of the current in the machine, indicating that the assumption for Eq. (4.7) is reasonable. Finally, Fig. 4.11d shows that most residual energy (2182 J) is dissipated by the BR and only 20.2 J is consumed by the machine windings. Before leaving Fig. 4.11, an interesting phenomenon that the output electromagnetic torque of the machine fluctuates greatly during discharge should not be ignored. This happens due to the following reasons. It needs to be mentioned that the torque is directly observed by (4.7), so the dynamics of the torque see the similar trend with the *q*-axis current. Notably, the torque experiences a step change immediately when a discharge request occurs. Meanwhile, when the *q*-axis current declines as the bus voltage decreases, the magnitude of the torque will gradually drop as well. Further, because the PMSM works as a three-phase generator and the inverter functions as an UR when the bleeder-based discharge mode is activated, the currents in each phase of the machine are not continuous (with fluctuations) due to the properties of an UR [129]. Even worse, there exist the moments at which only two-phase windings are conducted and the other phase is shut off in each electrical period. Consequently, when transforming the three-phase currents into the ones in the rotating reference

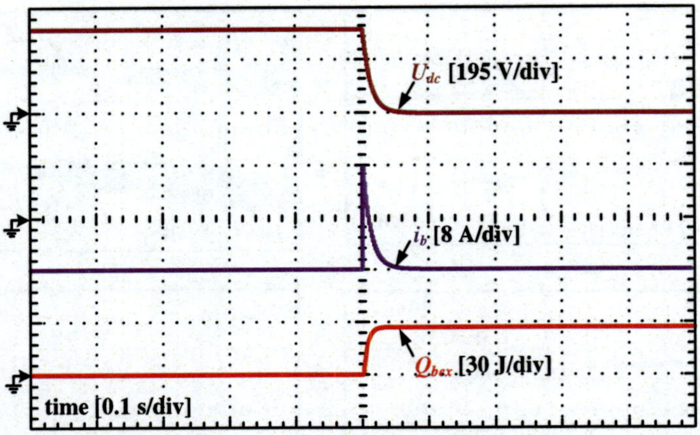

Fig. 4.12 Experimental results of the proposed discharge mode when emergency occurs at the speed of zero

frame, large fluctuations will be witnessed as in Fig. 4.11b. Further, the output braking torque will be influenced so as to be fluctuant as in Fig. 4.11c.

Given that the discharge is requested when the machine speed is zero, the experimental results are shown in Fig. 4.12. The bus voltage can drop to zero within about 0.03 s, which means the voltage balance time is really short and the safety can be ensured. All of the energy stored in the capacitor (about 27 J) will be consumed by the BR.

Overall, according to the experimental results, the proposed full-power discharge algorithm, the partial-power discharge algorithm and the bleeder-based discharge algorithm are proven to be able to achieve fast discharge within five seconds for the tested system, complying with the requirements in the UN ECE Regulation R94.

4.5 Summary

This chapter proposes a novel hybrid residual energy discharge strategy based on the internal windings and the external bleeder circuits to achieve the fast discharge requirement without greatly sacrificing the BR size and weight for high-voltage PMSM powertrain-based EVs during crash. Firstly, an accurate model for the bleeder-based discharge method is established to explicitly evaluate the BR size and weight sacrifice, explaining its defects when it is applied to a system with extreme parameters and "five-second discharge" is required. Secondly, the combined discharge strategy that synthesizes the advantages of the winding-based and external bleeder-based discharge schemes is detailed. It is proven that by using the hybrid discharge method, the size and weight of the BR decrease significantly compared to the traditional external bleeder-based discharge strategy. Moreover, because the proposed method can achieve the best (five-second) discharge process for any EV drives, it contributes to high crash safety with respect to electric shock prevention.

Chapter 5
Fault-Tolerant Winding-Based DC-Bus Capacitor Discharge Strategy

This chapter aims to tackle the challenge that a DC-bus capacitor discharge strategy which uses the machine windings as energy dissipation devices does not work when the rotor position is lost during crash. Specifically, efficient position loss fault-tolerant (PL-FT) control algorithms are designed and then incorporated into the winding-based discharge strategies to improve the reliability of the system. Firstly, an enhanced second order sliding mode (SOSM) observer with strong robustness against the DC-bus voltage disturbances is developed to detect the rotor position over the high-speed range. Secondly, an adaptive sine-wave (SW) based low-speed position observer (LSPO) with high immunity to voltage changes is designed for position estimation in the low-speed situations. By using the estimated rotor position, a winding-based discharge method is still effective even though the sensors for measuring rotor position malfunction during crash. Apart from endowing the winding-based discharge strategies with fault-tolerance capability, this chapter also puts forward new indirect current regulation-based discharge methods which are completely different from the aforementioned methods (traditional LDA-CI and NDNQ methods, etc.). Eventually, both simulation and experiment are carried out to verify the proposed fault-tolerant winding-based discharge techniques.

5.1 Introduction

The implementation of the discharge methods that function machine windings as the bleeder must use accurate rotor position which is usually provided by an external position sensor (eg., resolver) embedded in the machine. Although the sensor body is unlikely to fail during crash because of its robust construction and the relatively safe location in EVs, the transmission wires of position signals are inclined to come loose or fall out, resulting in the failure of position loss. A doable solution to this problem is to incorporate position sensor fault handling techniques (over full-speed

C. Gong, *Crash Safety of High-Voltage Powertrain Based Electric Vehicles*, Springer Theses, https://doi.org/10.1007/978-3-030-88979-1_5

range) into the discharge process. Nevertheless, another problem that how to design efficient and suitable PL-FT controllers arises.

In contrast with the normal operations, the DC-bus voltage cannot remain stable in the course of discharge after crash, with sharp decline and acute fluctuations. Thus, the PS-FT controllers must have robustness against DC-bus voltage variations [130], [131], otherwise the discharge performance will deteriorate or even worse, failure occurs. Among the traditional model-based PL-FT control methods for the high-speed situations, which include Luenberger observer [132], extended Kalman filter [133, 134] and sliding mode (SM) observer [135, 136], the last theory is well-known for its robustness [137, 138], so it can be adopted as the high-speed position observer (HSPO) for PL-FT control. However, the conventional SM observer has the disadvantage of chattering effect because signum function is used as the switching function [139], aggravating the inaccuracy of position estimation. Below are the explanations for the chattering effect. In the theoretical description of sliding modes, the system stays confined to the sliding surface and need only be viewed as sliding along the surface. However, real implementations of sliding mode control approximate this theoretical behavior with a high-frequency and generally non-deterministic switching control signal that causes the system to "chatter" in a tight neighborhood of the sliding surface. Chattering can be reduced through the use of dead bands or boundary layers around the sliding surfaces. For the standstill and low-speed conditions, high frequency (HF) voltage injection methods, including sine-wave voltage injection [140–142] and square-wave voltage injection [143–145], can be employed for PL-FT control. However, the induced HF currents that are used to obtain the position information becomes weaker as the DC-bus voltage declines, influencing the estimation precision. This undoubtedly poses a threat to the reliability of the winding-based discharge process. In summary, PL-FT control or position estimation techniques are mature for the PMSMs working normally, but there are few studies concerning the systems with variable DC-bus voltage.

Taking the above problems into account, this chapter introduces reliable winding-based DC-bus capacitor discharge techniques that have the fault-tolerance capability against position loss. In detail, an enhanced SOSM observer and an adaptive SW-LSPO are specially designed for PL-FT control during the discharge process, which can improve the system reliability. It deserves to be mentioned that the proposed PL-FT techniques are applicable to all of the aforementioned discharge strategies (traditional LDA-CI, NDNQ and hybrid methods, etc.) which need machine windings. However, in order to further enrich the winding-based discharge theories, new indirect current regulation-based discharge strategies are proposed in this chapter.

5.2 Design of HSPO Based on SM Theory

5.2.1 *Machine Modelling*

When designing a PL-FT control algorithm based on SM theory for the high-speed conditions, the PMSM model in the two-phase stationary (α, β) reference frame is needed. After applying the Inverse Park Transformation Eqs. (3.7) and (3.8), the electrical model of PMSM in the stationary reference frame can be expressed as the following state-space equations, where the iron saturation, magnetic flux leakage, eddy current and hysteresis loss are assumed to be negligible:

$$\begin{bmatrix} \frac{\mathrm{d}i_\alpha}{\mathrm{d}t} \\ \frac{\mathrm{d}i_\beta}{\mathrm{d}t} \end{bmatrix} = \begin{bmatrix} -\frac{R_s}{L_d} & -p\omega_m \frac{L_d - L_q}{L_d} \\ p\omega_m \frac{L_d - L_q}{L_d} & -\frac{R_s}{L_d} \end{bmatrix} \begin{bmatrix} i_\alpha \\ i_\beta \end{bmatrix} + \frac{1}{L_d} \begin{bmatrix} u_\alpha \\ u_\beta \end{bmatrix} - \frac{1}{L_d} \begin{bmatrix} e_\alpha \\ e_\beta \end{bmatrix} \tag{5.1}$$

The back EMF e_α, e_β of the machine can be represented as:

$$e_\alpha = -\left((L_d - L_q)\left(p\omega_m i_d - \frac{\mathrm{d}i_q}{\mathrm{d}t} \right) + p\omega_m \Psi_f \right) \sin\theta \tag{5.2}$$

$$e_\beta = \left((L_d - L_q)\left(p\omega_m i_d - \frac{\mathrm{d}i_q}{\mathrm{d}t} \right) + p\omega_m \Psi_f \right) \cos\theta \tag{5.3}$$

It can be noticed that the α, β-axis back EMFs contain the real rotor position θ, but they cannot be measured directly. On this ground a sliding mode variable structure controller will be employed to estimate the real-time back EMFs, from which the position information can be extracted.

5.2.2 *Traditional SOSM Observer*

Referring to [146], traditional SOSM observers used for estimating the back EMF of system (5.1) can be constructed as:

$$\begin{aligned} \frac{\mathrm{d}\hat{i}_\alpha}{\mathrm{d}t} =& -\frac{R_s}{L_d}\hat{i}_\alpha - p\omega_m \frac{L_d - L_q}{L_d}\hat{i}_\beta + \frac{1}{L_d}u_\alpha \\ &+ \frac{1}{L_d}\left(\lambda_1 |\bar{i}_\alpha|^{\frac{1}{2}} \mathrm{sign}(\bar{i}_\alpha) + \int_0^t (\lambda_2 \mathrm{sign}(\bar{i}_\alpha) + \rho) \right) \\ \frac{\mathrm{d}\hat{i}_\beta}{\mathrm{d}t} =& p\omega_m \frac{L_d - L_q}{L_d}\hat{i}_\alpha - \frac{R_s}{L_d}\hat{i}_\beta + \frac{1}{L_d}u_\beta \end{aligned} \tag{5.4}$$

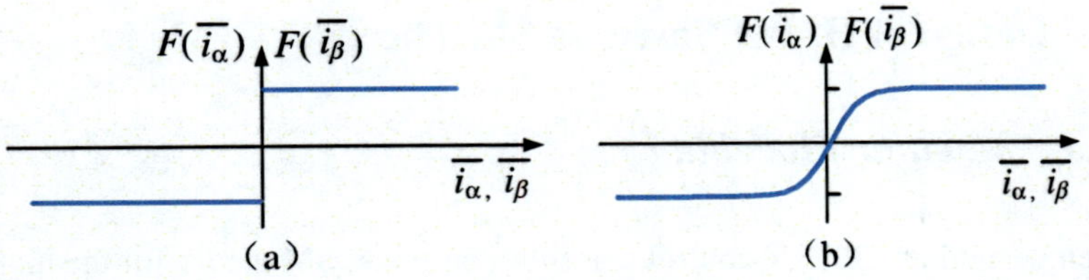

Fig. 5.1 Properties of switching function. **a** Signum function. **b** Sigmoid function

$$+\frac{1}{L_d}\left(\lambda_1|\bar{i}_\beta|^{\frac{1}{2}}\text{sign}(\bar{i}_\beta)+\int_0^t(\lambda_2\text{sign}(\bar{i}_\beta)+\rho)\right) \tag{5.5}$$

It can be seen that signum function (see Fig. 5.1a) is used as the switching function in the traditional SOSM observers. Besides, in order to ensure global stability, the perturbation term ρ can be designed as:

$$\rho = 0 \tag{5.6}$$

When (5.4) and (5.5) get to the equilibrium state in the control process, the estimated back EMFs used for position estimation are as follows:

$$\hat{e}_\alpha=\lambda_1|\bar{i}_\alpha|^{\frac{1}{2}}\text{sign}(\bar{i}_\alpha)+\int_0^t(\lambda_2\text{sign}(\bar{i}_\alpha)) \tag{5.7}$$

$$\hat{e}_\beta=\lambda_1|\bar{i}_\beta|^{\frac{1}{2}}\text{sign}(\bar{i}_\beta)+\int_0^t(\lambda_2\text{sign}(\bar{i}_\beta)) \tag{5.8}$$

The above equations illustrate that the chattering effects of an SOSM controller can be attenuated because the integral terms are able to prevent violent variations caused by signum function. However, the first terms in (5.7) and (5.8), which contribute to the dynamic characteristics still contain signum functions. Consequently, it is impossible to eliminate the chattering phenomenon totally. The experimental results in [147] have also proven that the chattering phenomenon will arise in the above traditional SOSM observer. In this case, it is preferable to employ low pass filters (LPF) to further attenuate the chattering effects, inevitably reducing the system bandwidth.

5.2.3 Proposed Enhanced SOSM Observer

This section will present an enhanced SOSM observer comprising a continuous sigmoid function (CSF) (replacing the signum function) as the switching function to further reduce the chattering effects (see Fig. 5.1b).

(a) **Observer design**

The sigmoid function $F(\bar{i}_{\alpha,\beta})$ is expressed as:

$$F(\bar{i}_\alpha) = \frac{2}{1+\exp(-\bar{i}_\alpha)} - 1, \quad F(\bar{i}_\beta) = \frac{2}{1+\exp(-\bar{i}_\beta)} - 1 \tag{5.9}$$

Substitute (5.9) for the signum functions in (5.4) and (5.5) and let the extra term ρ equal zero. A CSF-based SOSM observer can be rewritten as:

$$\begin{aligned}\frac{d\hat{i}_\alpha}{dt} =& -\frac{R_s}{L_d}\hat{i}_\alpha - p\omega_m\frac{L_d - L_q}{L_d}\hat{i}_\beta + \frac{1}{L_d}u_\alpha \\ &+ \frac{1}{L_d}\left(\lambda_1|\bar{i}_\alpha|^{\frac{1}{2}}F(\bar{i}_\alpha) + \int_0^t(\lambda_2 F(\bar{i}_\alpha))\right)\end{aligned} \tag{5.10}$$

$$\begin{aligned}\frac{d\hat{i}_\beta}{dt} =& p\omega_m\frac{L_d - L_q}{L_d}\hat{i}_\alpha - \frac{R_s}{L_d}\hat{i}_\beta + \frac{1}{L_d}u_\beta \\ &+ \frac{1}{L_d}\left(\lambda_1|\bar{i}_\beta|^{\frac{1}{2}}F(\bar{i}_\beta) + \int_0^t(\lambda_2 F(\bar{i}_\beta))\right)\end{aligned} \tag{5.11}$$

When the observer states reach the equilibrium point, the estimated back EMFs will become:

$$\hat{e}_\alpha = \lambda_1|\bar{i}_\alpha|^{\frac{1}{2}}F(\bar{i}_\alpha) + \int_0^t(\lambda_2 F(\bar{i}_\alpha)) \tag{5.12}$$

$$\hat{e}_\beta = \lambda_1|\bar{i}_\beta|^{\frac{1}{2}}F(\bar{i}_\beta) + \int_0^t(\lambda_2 F(\bar{i}_\alpha)) \tag{5.13}$$

By the use of CSFs, the terms that cause violent variations disappear, being able to reduce the chattering effects more effectively. Then, the estimated rotor position $\hat{\theta}$ can be calculated by:

$$\hat{\theta} = -\arctan\left(\frac{\hat{e}_\alpha}{\hat{e}_\beta}\right) \tag{5.14}$$

And the estimated angular speed $\hat{\omega}_m$ is equal to the differential of estimated position:

$$\hat{\omega}_m = \frac{\mathrm{d}\hat{\theta}}{\mathrm{d}t} \tag{5.15}$$

(b) **Stability analysis**

According to (5.1), (5.10) and (5.11), the model for current errors between the estimated values and the real values can be described as:

$$\begin{aligned}\frac{\mathrm{d}\bar{i}_\alpha}{\mathrm{d}t} = & -\frac{R_s}{L_d}\bar{i}_\alpha - p\omega_m\frac{L_d - L_q}{L_d}\bar{i}_\beta \\ & + \frac{1}{L_d}\left(e_\alpha - \lambda_1|\bar{i}_\alpha|^{\frac{1}{2}}F(\bar{i}_\alpha) - \int_0^t(\lambda_2 F(\bar{i}_\alpha))\right)\end{aligned} \tag{5.16}$$

$$\begin{aligned}\frac{\mathrm{d}\bar{i}_\beta}{\mathrm{d}t} = & p\omega_m\frac{L_d - L_q}{L_d}\bar{i}_\alpha - \frac{R_s}{L_d}\bar{i}_\beta \\ & + \frac{1}{L_d}\left(e_\beta - \lambda_1|\bar{i}_\beta|^{\frac{1}{2}}F(\bar{i}_\beta) - \int_0^t(\lambda_2 F(\bar{i}_\beta))\right)\end{aligned} \tag{5.17}$$

Define the sliding surfaces in the α, β axis as (S_α, S_β), and they can be denoted as:

$$\mathrm{S} = \begin{bmatrix} S_\alpha \\ S_\beta \end{bmatrix} = \begin{bmatrix} \bar{i}_\alpha \\ \bar{i}_\beta \end{bmatrix} \tag{5.18}$$

The Lyapunov function is constructed as:

$$V = \frac{1}{2}\cdot \mathrm{S}^T \cdot \mathrm{S} = \frac{1}{2}\bar{i}_\alpha^2 + \frac{1}{2}\bar{i}_\beta^2 \tag{5.19}$$

Obviously, $V > 0$. Then, according to the Lyapunov stability decision theorem, only by deducing that $\frac{\mathrm{d}V}{\mathrm{d}t} < 0$ will it can be concluded that the SMO can reach a stable state. Take the time derivative of Eq. (5.19):

$$\frac{\mathrm{d}V}{\mathrm{d}t} = \mathrm{S}^T \cdot \frac{\mathrm{dS}}{\mathrm{d}t} = \bar{i}_\alpha\frac{\mathrm{d}\bar{i}_\alpha}{\mathrm{d}t} + \bar{i}_\beta\frac{\mathrm{d}\bar{i}_\beta}{\mathrm{d}t} \tag{5.20}$$

Substitute (5.16) and (5.17) into (5.20), it can be further derived that:

$$\frac{dV}{dt} = \underbrace{-\left(\frac{R_s}{L_d}\bar{i}_\alpha^2 + \frac{R_s}{L_d}\bar{i}_\beta^2\right)}_{term1} \\ \underbrace{+\frac{\bar{i}_\alpha}{L_d}\left(_\alpha - \lambda_1|\bar{i}_\alpha|^{\frac{1}{2}}F(\bar{i}_\alpha) - \int_0^t(\lambda_2 F(\bar{i}_\alpha))\right)}_{term2} \\ \underbrace{+\frac{\bar{i}_\beta}{L_d}\left(e_\beta - \lambda_1|\bar{i}_\beta|^{\frac{1}{2}}F(\bar{i}_\beta) - \int_0^t(\lambda_2 F(\bar{i}_\beta))\right)}_{term3} \tag{5.21}$$

In order to keep the observer stable, *term*1, *term*2 and *term*3 are expected to be less than 0, so the parameters $\lambda_{1,2}$ should meet the following criteria:

$$\bar{i}_\alpha\left(_\alpha - \lambda_1|\bar{i}_\alpha|^{\frac{1}{2}}F(\bar{i}_\alpha) - \int_0^t(\lambda_2 F(\bar{i}_\alpha))\right)e < 0 \\ \bar{i}_\beta\left(e_\beta - \lambda_1|\bar{i}_\beta|^{\frac{1}{2}}F(\bar{i}_\beta) - \int_0^t(\lambda_2 F(\bar{i}_\beta))\right) < 0 \tag{5.22}$$

According to the sign (positive and negative) of $\bar{i}_{\alpha,\beta}$, Eq. (5.22) can be rewritten as:

$$\begin{cases} e_\alpha < \lambda_1|\bar{i}_\alpha|^{\frac{1}{2}}F(\bar{i}_\alpha) + \int_0^t(\lambda_2 F(\bar{i}_\alpha)), \text{ if } \bar{i}_\alpha > 0 \\ e_\alpha > \lambda_1|\bar{i}_\alpha|^{\frac{1}{2}}F(\bar{i}_\alpha) + \int_0^t(\lambda_2 F(\bar{i}_\alpha)), \text{ if } \bar{i}_\alpha < 0 \end{cases} \\ \begin{cases} e_\beta < \lambda_1|\bar{i}_\beta|^{\frac{1}{2}}F(\bar{i}_\beta) + \int_0^t(\lambda_2 F(\bar{i}_\beta)), \text{ if } \bar{i}_\beta > 0 \\ e_\beta > \lambda_1|\bar{i}_\beta|^{\frac{1}{2}}F(\bar{i}_\beta) + \int_0^t(\lambda_2 F(\bar{i}_\beta)), \text{ if } \bar{i}_\beta < 0 \end{cases} \tag{5.23}$$

Because the sign of $\bar{i}_{\alpha,\beta}$ is coincident with that of $F(\bar{i}_{\alpha,\beta})$, Eq. (5.24) is the strong condition for keeping the observer stable.

$$\lambda_1 > \max\left(\left|\frac{e_\alpha}{|\bar{i}_\alpha|^{\frac{1}{2}}F(\bar{i}_\alpha)}\right|, \left|\frac{e_\beta}{|\bar{i}_\beta|^{\frac{1}{2}}F(\bar{i}_\beta)}\right|\right), \quad \lambda_2 > 0 \tag{5.24}$$

In practice, $\bar{i}_\alpha$ and $\bar{i}_\beta$ are the current estimation errors. Assume that the observer can reach the equilibrium state and when requiring the observer remains stable with the error range of $[|\xi_{\min}|, |\xi_{\max}|]$, where $\xi_{\min}$ and $\xi_{\max}$ are the upper and lower limit of the current estimation errors, respectively, the minimum $|F(\bar{i}_{\alpha,\beta})|$ can be described as:

$$\min|F(\bar{i}_{\alpha,\beta})| = \frac{2}{1+\exp(-|\xi_{\min}|)} - 1 \tag{5.25}$$

Moreover, the magnitude of e_α and e_β cannot exceed the initial bus voltage U_{dc0} during discharge, that is,

$$e_\alpha < U_{dc0}\,,\ \ e_\beta < U_{dc0} \tag{5.26}$$

So far, it can be concluded that the enhanced observer will keep stable with the minimal estimation error of $|\xi_{\min}|$ when $\lambda_{1,2}$ satisfies the following conditions:

$$\lambda_1 > \frac{U_{dc}}{\sqrt{|\xi_{\min}|}\cdot\min|F(\bar{i}_{\alpha,\beta})|},\quad \lambda_2 > 0 \tag{5.27}$$

It can be noticed that the required $\xi_{\min}$ cannot be zero, which reflects the fact that the convergence property of a sigmoid function-based SM observer is sacrificed in comparison with the traditional signum function-based observer.

Finally, it should be addressed the crucial advantage of the proposed SOSM observer is that it is able to overcome the chattering effects totally without using LPFs. On this ground the time delay of the observer is small and the bandwidth for it is high. Consequently, the bandwidth and stability of the whole system are hardly influenced when the SOSM observer is adopted.

5.3 Design of Adaptive SW-LSPO

The DC-bus voltage descends continuously in the discharge process, but the traditional high frequency sine-wave voltage injection strategy used for estimating the rotor position relies on relatively high and stable DC-bus voltage. Thereby, a robust position observer based on adaptive principle against voltage drop is proposed in this section.

5.3.1 Traditional SW HF Injection Method

A derived expression of PMSM model used for position estimation over low-speed range is given as [14]:

$$\begin{bmatrix} u_\alpha \\ u_\beta \end{bmatrix} = R_s \cdot \begin{bmatrix} i_\alpha \\ i_\beta \end{bmatrix} + \begin{bmatrix} L_1 + L_2 \cos 2\theta & L_2 \sin 2\theta \\ L_2 \sin 2\theta & L_1 - L_2 \cos 2\theta \end{bmatrix} \times \frac{\mathrm{d}}{\mathrm{d}t} \begin{bmatrix} i_\alpha \\ i_\beta \end{bmatrix} + \frac{\mathrm{d}\theta}{\mathrm{d}t} \left(2L_2 \begin{bmatrix} -\sin 2\theta & \cos 2\theta \\ \cos 2\theta & \sin 2\theta \end{bmatrix} \begin{bmatrix} i_\alpha \\ i_\beta \end{bmatrix} + \Psi_f \begin{bmatrix} -\sin\theta \\ \cos\theta \end{bmatrix} \right) \tag{5.28}$$

where $L_1 = \frac{L_d + L_q}{2}$, $L_2 = \frac{L_d - L_q}{2}$. When the frequency of the injected voltage is high over the low-speed range, the impedance in the model acts more prominently than the reactance and the back EMF of the machine is low, so both the first and third terms in (5.28) can be neglected, namely:

$$u_\alpha = (L_1 + L_2 \cos 2\theta) \times \frac{\mathrm{d}i_\alpha}{\mathrm{d}t} + L_2 \sin 2\theta \times \frac{\mathrm{d}i_\beta}{\mathrm{d}t} \tag{5.29}$$

$$u_\beta = L_2 \sin 2\theta \times \frac{\mathrm{d}i_\alpha}{\mathrm{d}t} + (L_1 - L_2 \cos 2\theta) \times \frac{\mathrm{d}i_\beta}{\mathrm{d}t} \tag{5.30}$$

Traditionally, the HF voltages applied to the machine with constant amplitude U_h and angular frequency ω_h are expressed as follows:

$$u_{\alpha h} = U_h \cos(\omega_h t), \quad u_{\beta h} = U_h \sin(\omega_h t) \tag{5.31}$$

After the HF voltages are injected into the motor, the rotor position and speed can be calculated by synthetically processing (including phase shift, high pass filtering and phase lock loop control) the positive and negative sequences of the feedback HF currents in the stationary reference frame, which contain the position information:

$$i_{\alpha\beta h} = \overbrace{\frac{U_h(L_q + L_d)}{2\omega_h L_d L_q} e^{j(\omega_h t - \pi/2)}}^{\text{Postive sequence}} + \overbrace{\frac{U_h(L_q - L_d)}{2\omega_h L_d L_q} e^{j(-\omega_h t + 2\theta + \pi/2)}}^{\text{Negative sequence}} \tag{5.32}$$

5.3.2 *Impact of Bus Voltage on Sine-Wave HF Injection Method*

Many scholars have investigated the impact of amplitude of HF voltage on sensorless control, and it is found in [148] that the position error can be remarkably reduced as long as it rises to 20 V. Hence, in this study, the injected voltage amplitude is expected to be constant at that level during the whole discharge period. But it is not simple to achieve by means of traditional algorithms and the reasons are as follows.

In order to intuitively explain the problems, the below analysis will use normalized values of the voltage and current amplitude. Base voltage is defined in a rather conventional way, corresponding to the DC-bus voltage value in the normal situations,

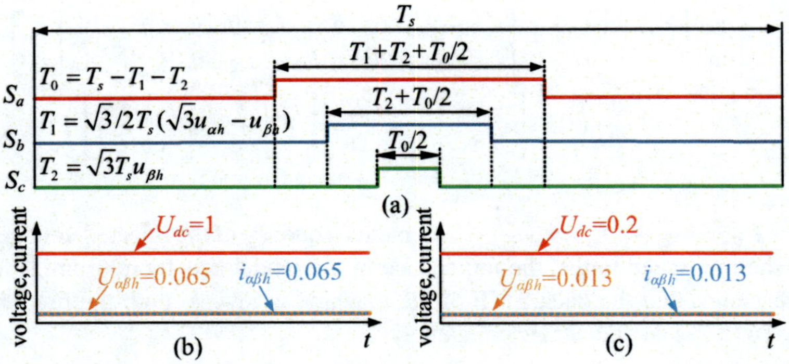

Fig. 5.2 **a** Switching time generated by HF voltage in the first sector. **b** Real HF voltage applied to the motor and feedback HF current when U_{dc} is 0.065. **c** Real HF voltage applied to the motor and feedback HF current when U_{dc} is 0.2

that is,

$$U_0 = U_{dc0} \tag{5.33}$$

Thus, the base voltage that equals rated value is calibrated to 1-p.u., at which 1-p.u. $\alpha\beta$-axis HF currents are induced. And 20 V is 0.065-p.u. in this study.

Take a motionless vehicle (standstill) as an example, U_h will be set to 0.065 at the beginning of discharge when the bus voltage U_{dc} is 1. At this moment, the feedback currents are 0.065. In this case, the switching time generated by HF voltage (e.g., in the first sector), real HF voltage applied to the motor and corresponding HF current are illustrated in Fig. 5.1a, b when space vector pulse width modulation (SVPWM) is employed. However, as is shown in (5.31), once the amplitude of the injected voltage is given, it no longer changes in the program as for the conventional sensorless algorithms, leading to unchanged calculated switching time like Fig. 5.2a. Hence, when U_{dc} declines, the real HF voltage applied to the machine cannot level off. Instead, it shows direct proportion with the DC-bus voltage. According to (5.32), the feedback currents will drop as well. Figure 5.2c shows the amplitude of the real HF voltage and current when U_{dc} is 0.2. In this case, the induced HF currents are weak, making it difficult to measure and process the useful signals. Because the signal-to-noise ratio is reduced, the accuracy of position estimation will degrade inevitably.

Overall, when the DC-bus voltage stands at a high position steadily, no problem is encountered when U_h is set to a constant small value. But as the DC-bus voltage is increasingly lower during discharge, the HF currents will get weaker and weaker over time. Therefore, the issue that the accuracy of position estimation decreases consistently arises.

5.3.3 *Proposed Adaptive SW-LSPO*

In order to maintain a constant level of the real HF voltage that is applied to the machine during discharge, an adaptive solution that compensates U_h in real time according to the DC-bus voltage downgrade is proposed. The implementation process can be divided into three different phases.

Phase 1: Detect the real-time bus voltage U_{dc} and calculate the drop-out value ΔU_{dc}.

$$\Delta U_{dc} = U_{dc0} - U_{dc} \tag{5.34}$$

Phase 2: Because the real injected HF voltage follows a linear relationship with both U_{dc} and switching time, the HF voltage loss caused by DC-bus voltage reduction can be compensated by extending the switching time generated by the set value of HF voltages in program (see Fig. 5.3). Obtain the compensation voltage:

$$\Delta U_h = k \Delta U_{dc} \tag{5.35}$$

where k is a correction coefficient, and it is 1 without considering the dead-time effects.

Phase 3: Set the HF voltage amplitude online in real time as:

$$U_{hr} = 0.065 + \Delta U_h \tag{5.36}$$

Theoretically, the increment of the given α, β-axis HF voltages $\Delta u_{\alpha h}$, $\Delta u_{\beta h}$ is equal to ΔU_h. At present, the real HF voltage applied to the machine is able to keep nearly 0.065-p.u., and the HF currents are produced accordingly during discharge, ensuring precise position estimation. Whereas, there exists another problem that the downward bus voltage will eventually arrive at the threshold 0.065. In this case, U_{dc} is not large enough to ensure HF signal injection and simultaneously maintain rotational motion. In this study, because the first priority is to discharge the capacitor voltage entirely, the rotor position will be estimated continuously following the

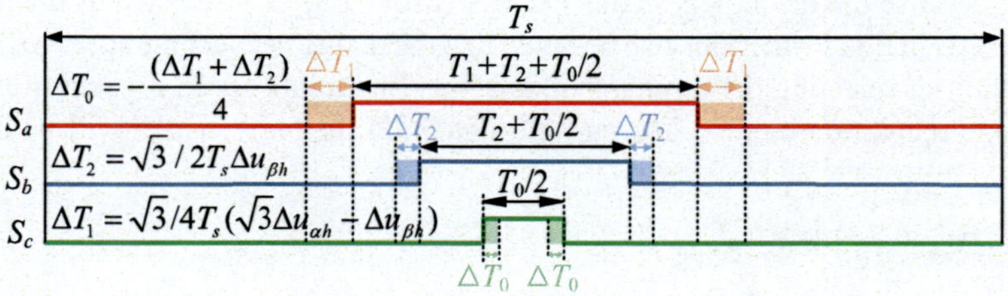

Fig. 5.3 Extended Switching time generated by HF voltage

above rules despite the degradation of accuracy. Although high discharge performance cannot be guaranteed, the residual energy is too small to cause harm over the period. It deserves to be that, firstly, the proposed position sensor fault-tolerant control algorithm is specially designed for the discharge process with declining DC-bus voltage. Secondly, the control performance of the machine will undoubtedly decrease (torque and speed ripples increase) because the HF voltage and current are injected into the machine. This is acceptable for the winding-based discharge process because high-control performance is not necessary, and what needs to be focused on is the function of high-voltage discharge. Hence, in this research, the impacts of the HF signals on the machine control performance (torque and speed ripples etc.) can be ignored.

5.4 Fault-Tolerant Full-Speed Range Discharge

When integrating the HSPO and LSPO into the discharge process, an EV powertrain with modest parameters will be considered in this section. In terms of this kind of system, the traditional LDA-CI discharge method is effective to achieve the best discharge process. However, in order to enrich the winding-based discharge theories, this section does not just use the traditional LDA-CI method, in which the d, q-axis currents are directly set as the targeting control objectives (direct current regulation-based discharge algorithm). Supplementarily, indirect current regulation-based discharge algorithms based on voltage and speed regulation are developed for different working states.

The flow diagram of the proposed fault-tolerant winding-based discharge algorithms is shown in Fig. 5.4a. Once a discharge request occurs, the instantaneous feedback rotor speed ω_{m0} will be functioned as the selection signal for the deceleration process or the acceleration process. Then, in the discharge process, the real-time U_{dc} determines whether the discharge is supposed to stop. During deceleration, ω_m is used for selecting the position observer. Since the HSPO and LSPO are suited for different speed range, a switching point $\omega_{pos.th}$ needs to be set beforehand for the sake of normal implementation of the PL-FT discharge scheme. Usually, $\omega_{pos.th}$ can be set as a value that is lower than $\omega_{m.th}$ for a PMSM used in EVs. During acceleration, because the machine speed is extremely low, LSPO is selected as the position observer. In Fig. 5.4a, it can also be noted that the real-time machine speed ω_m has the function of selecting the discharge algorithms (including speed regulation method, voltage regulation method and current regulation method), which will be detailed next.

(a) Deceleration process

The deceleration process comprises two steps (voltage and current regulation), as in Fig. 5.4b. In order to explain it roundly, take that ω_{m0} > as an example and both the control methods will be experienced.

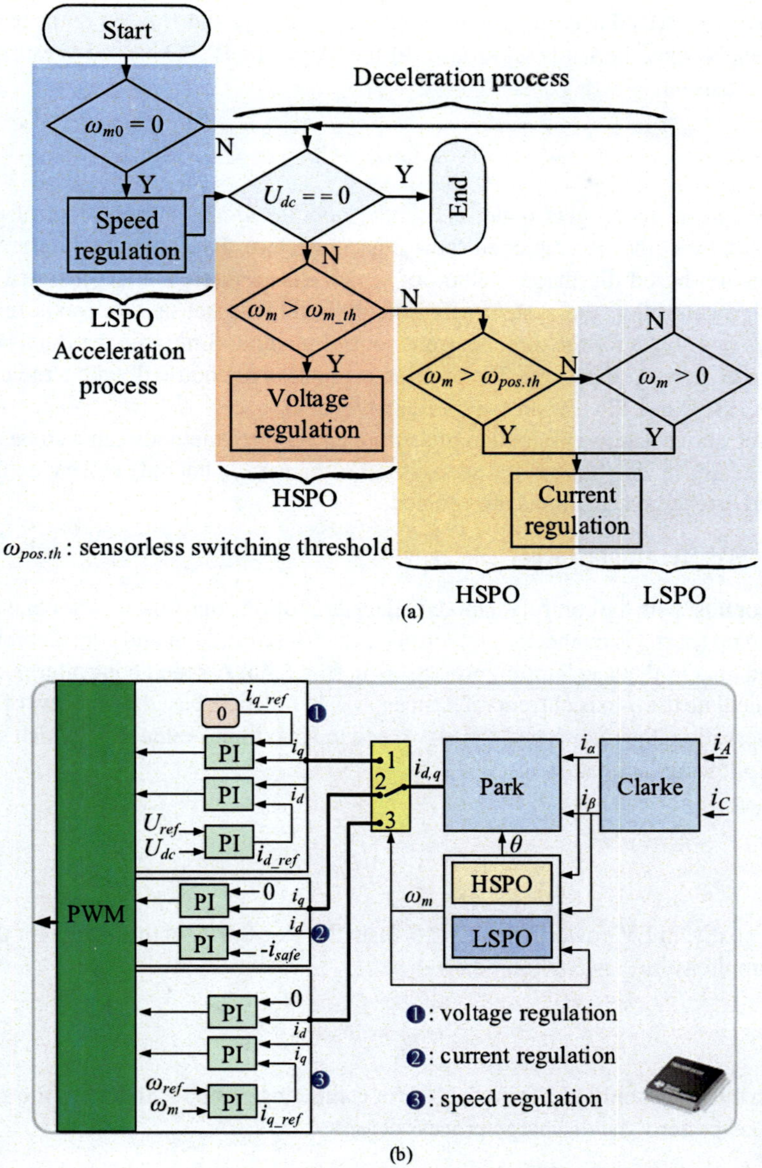

Fig. 5.4 Winding-based discharge scheme. **a** Flow diagram. **b** Control methods

When the rotor speed is higher than ω_{m_th}, a DC-bus voltage regulation algorithm is activated (port 1 is connected) to maintain the bus voltage at five percent lower (U_{ref}) than the safe voltage until the machine speed decreases under ω_{m_th}. In detail, a voltage proportion-integral (PI) controller is adopted to produce a large negative d-axis current reference i_{d_ref} which can regulate the magnitude of the air-gap magnetic field of the PMSM according to the error between U_{ref} and U_{dc}. As far as the q-axis current is concerned, it is set to zero. At this stage, the HSPO is used to estimate the rotor position.

$$i_{q\_ref} = 0 \tag{5.37}$$

When the speed gets under ω_{m_th}, the direct d, q-axis current control method (LDA-CI method) is employed (port 2 is connected). By contrast with the voltage regulation-based discharge method of which d-axis reference is generated by the voltage controller, the system safe current is directly set as the d-axis reference ($i_{d\_ref} = -I_{safe}$) for the direct current control method. Similarly, the q-axis current keeps at zero. At this stage, the position estimation method will take a radical shift from HSPO to LSPO when the speed arrives at $\omega_{pos.th}$.

The above energy dissipation procedure and control methods can also be applied to the cases in which the initial speed is lower than ω_{m_th}, but only current regulation-based discharge algorithms are enabled.

(b) **Acceleration process**

By contrast with the control methods in deceleration, the machine will be controlled to speed up (port 3 is connected) by setting a zero d-axis current and constant reference speed ω_{ref} in the acceleration process as in Fig. 5.4b. A speed controller is utilized to calculate the q-axis current reference i_{q_ref}. In practice, based on the law of energy conservation, the rotor speed cannot exceed the maximum value $\omega_{\max}$ which satisfies the following inequation:

$$\omega_{\max} < \sqrt{\frac{C}{J}} \cdot U_{dc0} \tag{5.38}$$

Usually, in EV powertrains, $C \ll J$ (usually $\frac{C}{J} < 0.01$). In this study, for the sake of simplicity, ω_{ref} is configured as:

$$\omega_{ref} = \omega_{pos\_th} \tag{5.39}$$

In this stage, only LSPO is enabled for estimating position, and when the detected U_{dc} keeps zero, the discharge process ceases.

(c) **Design and Analysis for Voltage Controller (PI)**

In Fig. 5.4b, compared to the speed and current PI controllers that have been widely used in the PMSM drive systems, the voltage PI controller is seldom adopted so that there is a lack of design technique for it. However, because it is an important part of the discharge technique, it is necessary to develop the standard parameter tuning and analysis method for the regulator.

The structure of the voltage regulation-based discharge control system is shown in Fig. 5.5. It can be seen that the voltage controller is only embedded in the d-

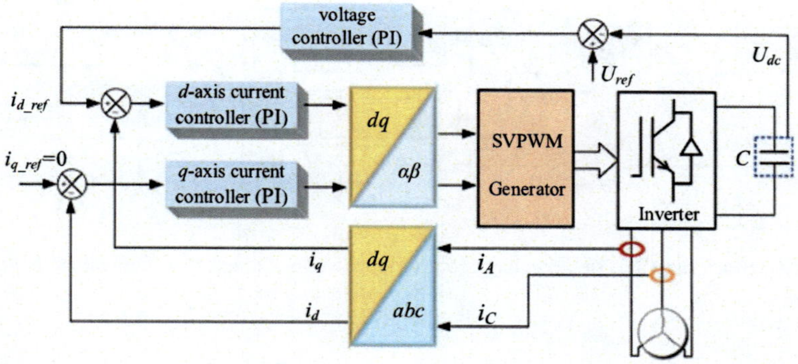

Fig. 5.5 Structure of the voltage regulation-based discharge control system

axis current control loop of the machine, so the d-axis electrical dynamics of the PMSM are needed for designing the parameters of voltage PI controller, which can be described as:

$$\frac{di_d}{dt} = -\frac{R_s}{L_d}i_d + \frac{L_q}{L_d}p\omega_m i_q + \frac{u_d}{L_d} \tag{5.40}$$

Moreover, the relationship between the DC-bus voltage and the d-axis current is needed [with reference to (3.10)]:

$$U_{dc} = \sqrt{3}C_e'(\Psi_f + L_d i_d)\omega_m \tag{5.41}$$

In addition, assume that the motor drive control algorithm allows perfect decoupling between the d and q axis [149], and then, the machine model in the s-domain can be obtained by applying Laplace transform to (5.40) and (5.41):

$$G_c(s) = \frac{i_d(s)}{u_d(s)} = \frac{1}{sL_d + R_s} \tag{5.42}$$

$$U_{dc}(s) = \frac{\sqrt{3}C_e'\omega_m\Psi_f}{s} + \sqrt{3}C_e'\omega_m L_d i_d(s) \tag{5.43}$$

Then, the inverter used to supply the machine would be approximately modeled as a simple gain 1 when the space vector pulse width modulation (SVPWM) strategy is adopted, that is,

$$G_{inv}(s) = 1 \tag{5.44}$$

As for the voltage and current PI controllers, they can be described as follows:

$$G_{u\_pi}(s) = k_{up} + \frac{k_{ui}}{s} \tag{5.45}$$

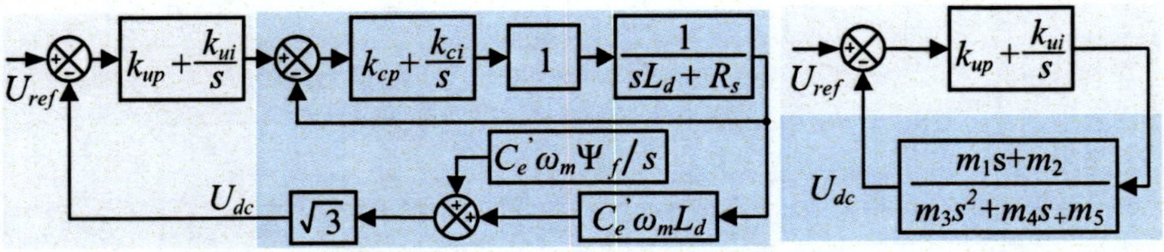

Fig. 5.6 Block diagrams of the voltage control loop. **a** Block diagram. **b** Equivalent diagram

$$G_{c\_pi}(s) = k_{cp} + \frac{k_{ci}}{s} \tag{5.46}$$

where k_{up} and k_{ui} are the proportional and integral gain of the voltage PI controller, respectively. The proportional and integral gain of the current PI controller is k_{cp} and k_{ci}, and they can be calculated following the traditional method in [150].

Then, the block diagram of the voltage and current control loop in the *s*-domain can be depicted as in Fig. 5.6a. For a fixed system, the parameters of the machine and current controller are known. For the sake of simplicity, the system can be regarded as one second-order system (see Fig. 5.6b) and the transfer function is:

$$G_{equ}(s) = \frac{m_1 s + m_2}{m_3 s^2 + m_4 s + m_5} \tag{5.47}$$

where $m_1 = \sqrt{3}C_e'\omega_m L_d k_{cp}$, $m_2 = \sqrt{3}C_e'\omega_m L_d k_{ci}$, $m_3 = L_d$, $m_4 = R_s + k_{cp}$, $m_5 = k_{ci}$. Then, the closed-loop system transfer function can be described as:

$$C(s) = \frac{k_{up}m_1 s^2 + (k_{up}m_2 + k_{ui}m_1)s + k_{ui}m_2}{m_3 s^3 + (m_4 + k_{up}m_1)s^2 + (m_5 + k_{up}m_2 + k_{ui}m_1)s + k_{ui}m_2} \tag{5.48}$$

Meanwhile, the characteristic equation can be obtained:

$$D(s) = m_3 s^3 + (m_4 + k_{up}m_1)s^2 + (m_5 + k_{up}m_2 + k_{ui}m_1)s + k_{ui}m_2 \tag{5.49}$$

Take the drive system with the parameters in Table 5.1 ($k_{cp} = 3.8$, $k_{ci} = 0.5$) as an example, the Routh criterion should be employed to analyze the system stability and design the system parameters [151], and the Routh table is shown in Table 5.2. In order to keep the system stable, the following conditions need to be satisfied:

$$k_{up} > 0, k_{ui} > 0, \omega_m < 1845\,\text{rad/s} \tag{5.50}$$

Because the tested machine always operates under 1845 rad/s, it is simple to determine the values of k_{up} and k_{ui} following (5.50). Further, in order to ensure the rapid response speed (fast voltage regulation), the bandwidth of the system should

Table 5.1 Parameters of PMSM powertrain

Parameter	Value	Unit
Stator winding resistance R_s	0.18	Ω
d-axis inductance L_d	4.2	mH
q-axis inductance L_q	10.1	mH
Number of pole pairs p	3	–
Moment of inertia J	0.052	kg m^2
Sensorless switching speed $\omega_{pos.th}$	30	rad/s
Threshold speed of safety ω_{m_th}	60	rad/s
Rated speed ω_{rated}	314	rad/s
DC-bus voltage U_{dc0}	310	V
System safe current I_{safe}	100	A
Voltage constant C'_e	3.16	–
DC-bus capacitor C	560	μF

Table 5.2 Routh table

s^3	0.0042	$0.5 + 0.011\omega_m k_{up} + 0.087k_{ui}$
s^2	$0.98 + 0.087k_{up}$	$0.011\omega_m k_{ui}$
s^1	$(0.98 + 0.087k_{up})(0.5 + 0.011\omega_m k_{up}) + \dfrac{7.65e^{-3}k_{ui}k_{up} + (0.08526 - 4.62e^{-5}\omega_m)k_{ui}}{0.98 + 0.087k_{up}}$	0
s^0	$0.011\omega_m k_{ui}$	0

not be reduced significantly after using the PI controller. In this case, both the gain k_{up} and the shear frequency $\omega_{u\_c} = \frac{k_{ui}}{k_{up}}$ of the voltage PI controller should stand at a relatively high position according to the automatic control theory. In practice, the bandwidth loss over the full-speed operating range is little once the following conditions are satisfied:

$$k_{up} > 1, \quad \frac{k_{ui}}{k_{up}} > p\omega_{rated} \tag{5.51}$$

Finally, according to the above analysis, it needs to be explained that the reason why the author calls the voltage regulation- and speed regulation-based discharge methods as indirect current regulation-based methods is that although the direct targeting control objectives are voltage and speed, respectively, in essence, the discharge process is achieved by injecting particular currents (current references are generated by voltage and speed controllers).

5.5 Simulation and Experimental Verifications

5.5.1 Simulation Results

To verify the proposed discharge methods comprehensively, both high-speed and low-speed (including standstill) cases are considered. In these cases, the deceleration and acceleration algorithms can be verified.

(a) **Results of deceleration algorithms**

In order to verify the fault-tolerant discharge strategy, the PMSM drive for EV with parameters in Table 5.1 is studied. Figure 5.7a shows the fault-tolerant discharge characteristics when discharge request occurs at the speed of rated value ω_{rated}. When the voltage regulation algorithm launches, the capacitor voltage drops under 60 V sharply. This happens because nearly − 75 A d-axis current is generated, weakening the air-gap field of the motor instantly. At the moment, the DC-bus capacitor voltage is much higher than the back EMF, so the energy can flow fast from the capacitor to PMSM. But soon the capacitor voltage will not be constantly higher than the back EMF due to the gradually declining rotor speed. Afterwards, the capacitor will be in a status of discharge accompanied by charge. In the period, i_d keeps negative and a gradually increased q-axis current is produced due to cross-coupling effects (like the phenomenon in Chap. 3, Sect. 3.3.1) to keep the bus voltage at 57 V (with fluctuations) until about 3.3 s. Then, the voltage quickly decreases to zero because of larger i_d is seen immediately when the current regulation algorithm is adopted. Otherwise, the rotor speed cannot shift synchronously, leading to non-zero capacitor voltage and relatively low d, q-axis currents which do not track the prescribed trajectory between 3.3 and 4.0 s. The rotor position estimated by the proposed HSPO and the transient estimation error between 0.5 and 0.53 s are illustrated in Fig. 5.7b. It can be noticed that the enhanced SOSM observer can track the actual position with small error that is within ± 0.2 rad regardless of the bus voltage fluctuations, indicating that the proposed position observer is highly robust against the voltage variations so as to be suitable for the discharge process. However, Fig. 5.7c shows that the position estimation accuracy of the LSPO is deteriorating as the voltage and speed become lower. This happens because U_{dc} is not large enough to guarantee the signal-to-noise ratio of the HF voltage injection method and the maximum error reaches over 0.35 rad between 3.55 and 3.7 s.

Figure 5.8 demonstrates the discharge process which starts from ω_{m_th}, and the entire dissipation time is 0.67 s. At the start of the current regulation, a large negative d-axis current (but not as desired) is generated to drop the bus voltage violently to zero. At the moment, i_q experiences fluctuations arising from cross-coupling effects although the reference is set to zero. Comparatively, the speed witnesses slower decrease because of the relatively high mechanical time constant. Therefore, the capacitor voltage can still be slightly higher (fluctuations) than the final value and, both the i_d and i_q experience a wind-down period uncontrollably by the end of discharge. Figure 5.8b illustrates the position estimation results between 0.25 and

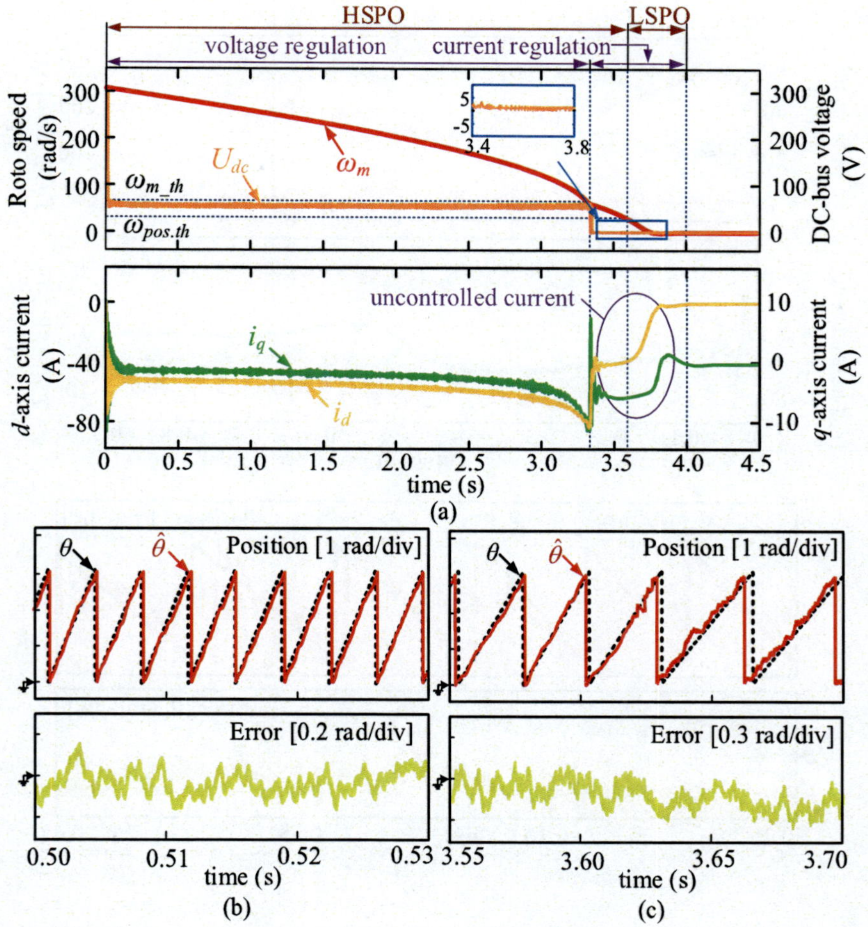

Fig. 5.7 Discharge process when discharge request occurs at rated speed. **a** Discharge characteristics. **b** Position estimation results at high speed and high voltage. **c** Position estimation results at low speed and low voltage

0.4 s. It can be seen that the estimated results of the position observers at low speed and voltage are not commendable. The maximum estimation error is analogously 0.45 rad, which is similar to that in Fig. 5.7c.

Before leaving Figs. 5.7 and 5.8, it can be noted that although the rotor position can be estimated by the proposed position sensor fault-tolerant algorithms over the full speed range and the deceleration discharge algorithms can be successfully implemented, both the current and position experience abrupt changes when the machine speed arrives at $\omega_{pos.th}$. This phenomenon is caused by the radical shift mode from HSPO to LSPO. One of the side effects of the unsmooth transition is that the discharge performance degradation problem occurs. Possibly, an overcurrent fault might arise,

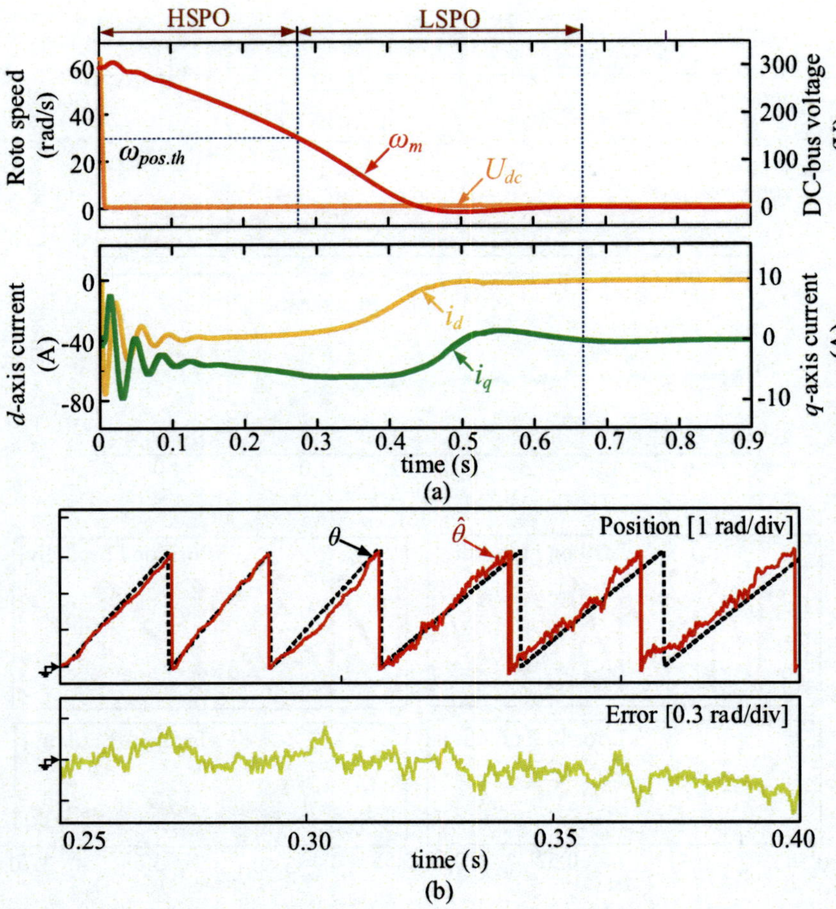

Fig. 5.8 Discharge process when discharge is requested at safety threshold. **a** Discharge characteristics. **b** Position estimation results

triggering the protection mode and interrupting the discharge process. Another issue that the vibration of the machine would intensify during the transition process arises because of the current fluctuations. Luckily, the transition happens when the bus voltage is below the safe level, so the electrical shock risks no longer exist. But in order to achieve higher reliability, the smooth transition methods from HSPO to LSPO still need to be developed.

(b) **Results of acceleration algorithms**

Assume that the emergency occurs at a standstill state and the acceleration process is tested as in Fig. 5.9. In this case, only LSPO (no transition) is used for detecting the position. Once a discharge request occurs, a positive q-axis current is generated to speed up the machine to about 6 rad/s, while the bus voltage jumped down to

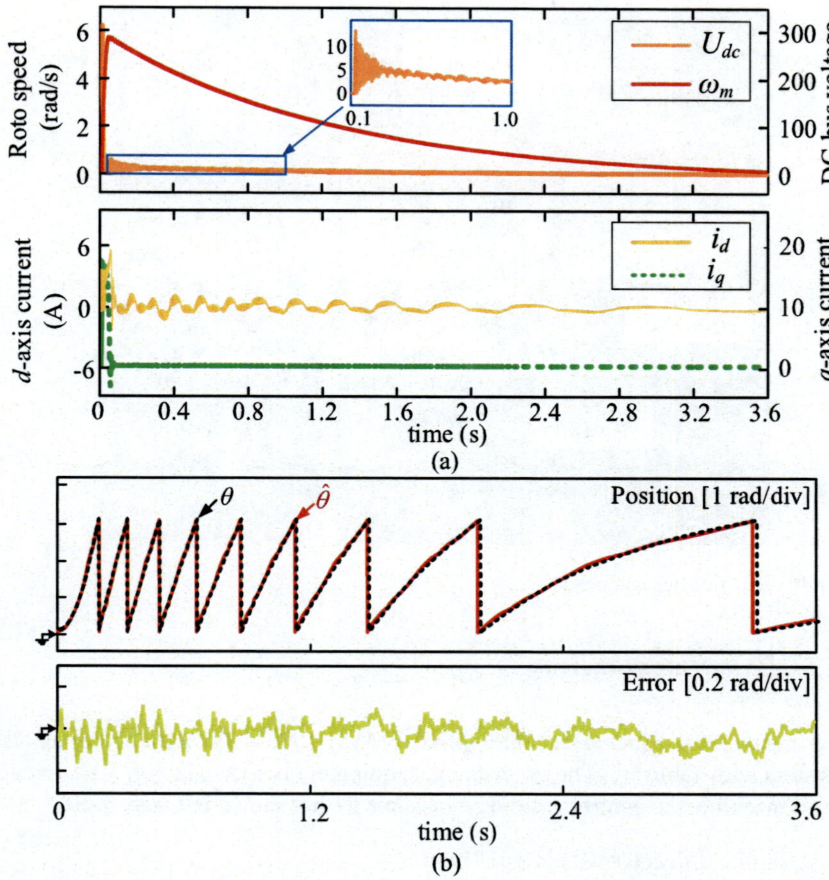

Fig. 5.9 Discharge process when discharge is requested at zero speed. **a** Discharge characteristics. **b** Position estimation results

an average of 5 V which is safe. Then, the machine slows down mainly relying on the friction from 0.4 to 3.6 s, while the windings can also expend energy due to the fluctuant *d*, *q*-axis currents. In terms of the position in Fig. 5.9b, interestingly, when the voltage is small, the estimated accuracy becomes higher than that in Fig. 5.7c and Fig. 5.8c. The reason why this happens is that the bus voltage for the acceleration-based algorithm declines gradually (see Fig. 5.9a), following the trend of speed. But the voltage just keeps jumping up and down around zero for the deceleration process in which the flux-weakening *d*-axis current and the negative *q*-axis current play a mutual role (see Fig. 5.7a).

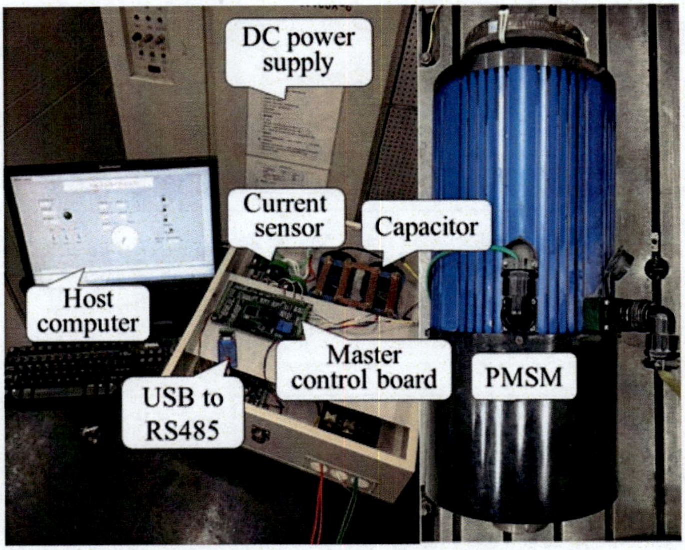

Fig. 5.10 Experimental equipment

5.5.2 *Experimental Results*

Experiments are conducted on a three-phase PMSM drive whose parameters are also consistent with Table 5.1. The experimental equipment is shown in Fig. 5.10. Except the test machine, the control system is the same to that shown in Chap. 3, Sect. 3.4.

(a) **Results of deceleration algorithms**

Assume a discharge request occurs when the motor speed is ω_{rated}. In accordance with the simulation results, Fig. 5.11a illustrates that the capacitor voltage drops under the safe level quickly at first and then levels off at the required position until 2.5 s when the proposed voltage regulation algorithm is employed. Once the currents in the machine are regulated, the bus voltage declines quickly to around zero. Figure 5.11b, d show the characteristics of the three phase currents and position information from 0.52 to 0.54 s in the voltage regulation range (Fig. 5.12 shows the current regulation case more clearly). It can be seen that the HSPO based on enhanced SOSM observer is able to track the real position with small error that is within $\pm$ 0.16 rad. The *d*, *q*-axis currents in Fig. 5.11c are very similar with those in Fig. 5.7. After discharge begins, i_d turns about − 60 A while only about − 2 A *q*-axis current is produced, and both of them show a slightly decreasing trend before the current regulation launches. Between 2.85 and 3.2 s, uncontrolled currents are generated as well, after which they go down to zero. Finally, the discharge time is about 3.0 s.

Figure 5.12 depicts the experimental results when discharge occurs at the speed of 60 rad/s and it can be noticed that the overall discharge time is about 0.6 s. In the discharge process, only current regulation algorithm is used. After the *d*- and *q*-axis

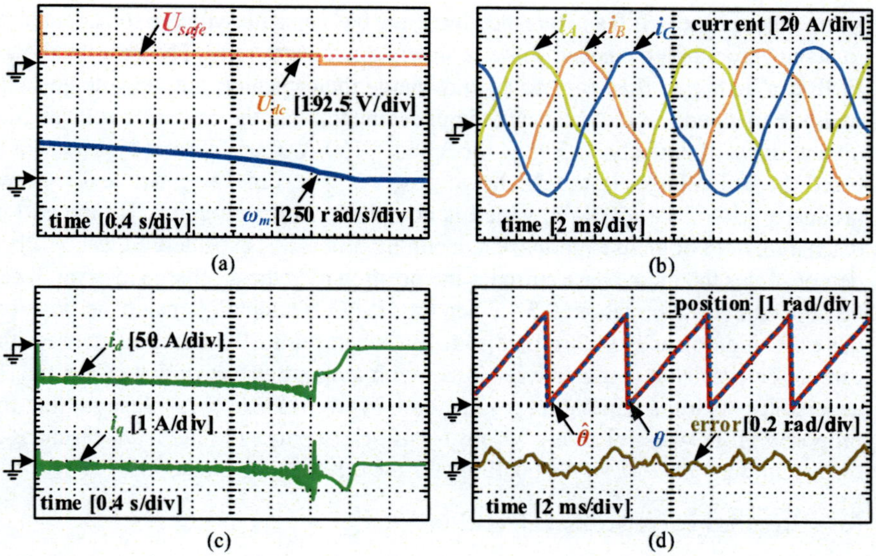

Fig. 5.11 Experimental results at the initial speed of 314 rad/s. **a** DC-bus capacitor voltage and machine speed. **b** Phase currents. **c** d, q-axis current. **d** Position information at high speed

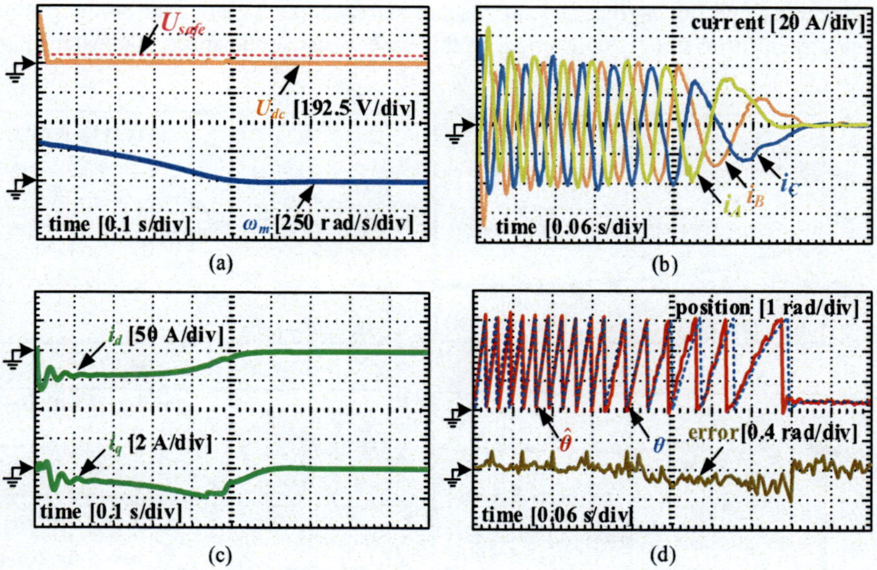

Fig. 5.12 Experimental results at the initial speed of 60 rad/s. **a** DC-bus capacitor voltage and machine speed. **b** Phase currents. **c** d, q-axis current. **d** Position information at low speed

references are set − 100 and 0 respectively, the bus capacitor voltage soon decreases to about zero while the speed shows a slowly downward trend. The generated back EMF can be higher than zero so as to recharge the capacitor, but the voltage surge will be expended instantly, causing the phenomenon that the capacitor voltage goes up and down alternately. In Fig. 5.12c, about − 50 A *d*-axis current is produced to bleed the bus voltage. In the discharge process, i_q becomes negative and it grows gradually to − 2 A at 0.44 s, which is compliant with simulation. By the end of discharge, both of them experience a declining process as expected. Figure 5.12b, d demonstrates the three phase currents and position over the discharge interval. It can be noticed that before about 0.3 s when the HSPO is adopted to calculate position, the estimation error is within 0.25 rad. But when the LSPO observer is used, the estimation precision gets relatively lower with a maximum deviation of 0.4 rad. In accordance with the simulation results, both Figs. 5.11 and 5.12 illustrate that the proposed HSPO has relatively strong robustness against the bus voltage changes (drop and fluctuations), so it's qualified for the discharge applications.

(b) **Results of acceleration algorithms**

Given that the discharge is requested when the machine speed is zero, the experimental results for the acceleration process are shown in Fig. 5.13. The bus voltage can drop under 60 V within 0.1 s as the rotor speed rises to about 5 rad/s. Then the machine slows down gradually and reaches zero at about 3 s when the entire residual energy in the system is consumed. In this case, because the speed is constantly set

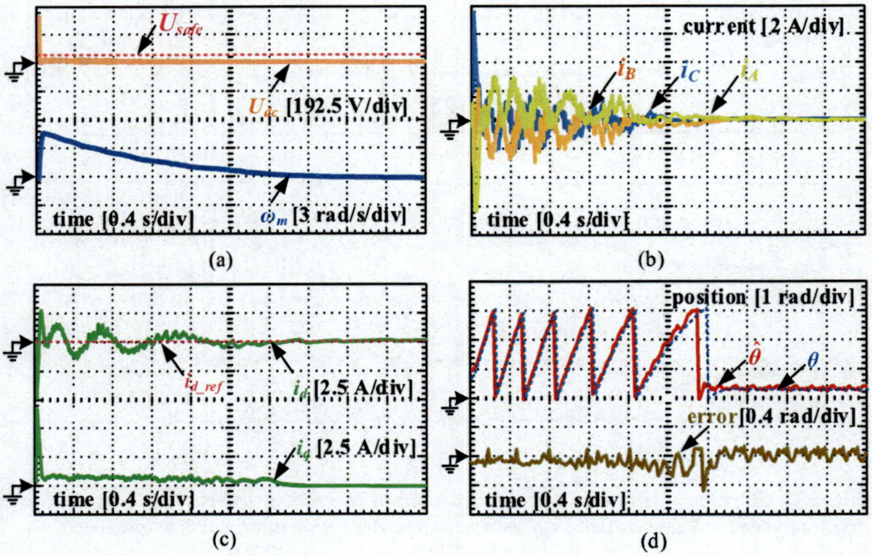

Fig. 5.13 Experimental results from standstill. **a** DC-bus capacitor voltage and machine speed. **b** Phase currents. **c** *d*, *q*-axis current. **d** Position information

as ω_{pos_th}, the q-axis current will keep positive while i_d experiences obvious fluctuations in Fig. 5.13c. As far as the position is concerned, overall, the proposed LSPO can track the real position with small error. Only when the speed approaches zero, a maximum estimation error of 0.4 rad is seen, which does not affect the discharge process any longer.

5.6 Summary

This chapter proposes effective PL-FT winding-based DC-bus capacitor discharge strategies to improve the system reliability. In order to handle the position loss faults, a robust HSPO based on enhanced SOSM theory and an adaptive SW-LSPO with high immunity to voltage changes are specially designed for position estimation in the high-speed and low-speed (including standstill) situations during crash, respectively. When implementing the winding-based discharge strategies, the rotor position information estimated by the position observers is employed, leading to the fact that the position loss fault does not influence the discharge process at all. In addition to the traditional LDA-CI discharge method, which is a direct current-regulation based strategy, a voltage regulation-based discharge method and a speed regulation-based discharge method are developed for the high-speed and standstill situations, respectively. By improving the reliability of the DC-bus capacitor discharge schemes, the crash safety level of EVs can be improved simultaneously.

Chapter 6
Winding-Based Discharge Technique Selection Rules Based on Parametric Analysis

In the previous chapters, several winding-based discharge strategies have been introduced, which include the traditional LDA-CI method, piecewise NDNQ method, hybrid discharge scheme and indirect current regulation-based methods. However, a crucial problem is that when designing the discharge system for a particular EV, there are no ready-to-use rules that an engineer can follow to pre-evaluate whether or which winding-based discharge technique is available, especially for the high-speed range. To solve this challenge, this chapter proposes general principles for picking out applicable winding-based DC-bus capacitor discharge techniques at the stage of EV design on the basis of parametric analysis. Then, case studies are used to verify the proposed discharge technique selection rules.

6.1 Introduction

From Chaps. 2 to 5, several winding-based discharge strategies have been introduced. In detail, Chaps. 2 and 3 illustrate that the traditional LDA-CI method is able to make the DC-bus voltage get down to a pretty low level instantaneously relying on the flux-weakening effect. A novel piecewise NDNQ discharge method is developed for the powertrains with extreme parameters in Chap. 3. In Chap. 4, a hybrid winding-based discharge method is designed for the systems with highly extreme parameters. Chapter 5 indicates that apart from the direct current-regulation based strategies, the DC-bus capacitor can be discharged by the indirect current regulation-based methods (voltage regulation- and speed regulation-based methods). The features of these winding-based discharge methods are summarized in Table 6.1. According to the features, a general classification method for the winding-based discharge techniques is given in Fig. 6.1. Before leaving Table 6.1, it can be noted that the pure winding-based discharge techniques that can be used for the high-speed situations could also be categorized into two groups: non-zero *d*-axis, zero *q*-axis (NDZQ) current

C. Gong, *Crash Safety of High-Voltage Powertrain Based Electric Vehicles*, Springer Theses, https://doi.org/10.1007/978-3-030-88979-1_6

Table 6.1 Characteristics of winding-based discharge strategies

Strategies	Typical features	Applications
Traditional LDA-CI method	(1) NDZQ currents (2) Instant discharge, but long-cycle discharge is accepted (3) Parameter-dependent (high-speed range) (4) Pure windings (5) Direct current regulation	(1) Running situations, (2) Standstill situations
Piecewise NDNQ method	(1) NDNQ currents (2) Long-cycle discharge (3) Parameter-dependent (high-speed range) (4) Pure windings (5) Direct current regulation	(1) High-speed situations
Voltage regulation-based method	(1) NDZQ currents (2) Instant discharge, but long-cycle discharge is accepted (3) Parameter-dependent (high-speed range) (4) Pure windings (5) Indirect current regulation	(1) Running situations, (2) Standstill situations
Speed regulation-based method	(1) ZDNQ currents (2) Pure windings (3) Indirect current regulation	(1) Standstill situations
Hybrid method	(1) External bleeder circuits and machine windings (2) Long-cycle discharge (3) Parameter-independent	(1) Running situations (2) Standstill situations

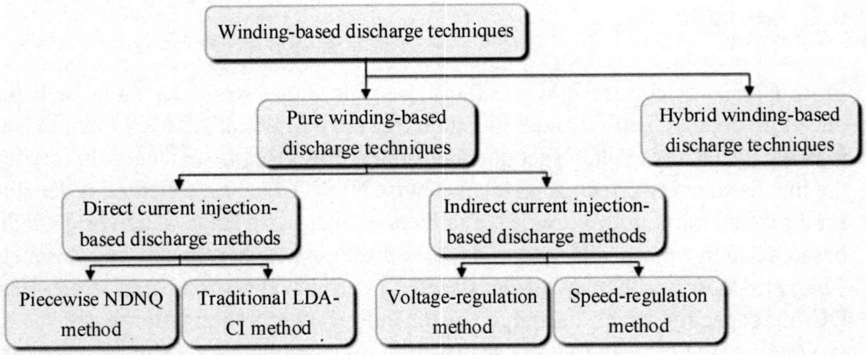

Fig. 6.1 Classification of winding-based discharge techniques

injection-based methods (including the traditional LDA-CI and voltage regulation-based methods) and NDNQ method. Since there are many kinds of winding-based discharge methods which have superior advantages (see Chap. 2), it is valuable to design a discharge system by using them for an EV in engineering.

When designing a discharge system for a new EV, one or more types of winding-based discharge methods can be employed as long as it/they are effective to achieve the fast discharge process. For the standstill cases (EV is hit by another car when it parks), because there is not much residual energy stored in the powertrain after crash, it is not difficult to select an appropriate winding-based discharge method. Table 6.1 shows that the traditional LDA-CI, voltage regulation-based and speed regulation-based discharge algorithms are available for the standstill cases. However, for the running situations (EV crashes when it runs on the road), it is possible that there is a huge amount of residual energy that needs to be dissipated (related to the initial speed and rotor inertia, etc. during crash). In this case, the selection process for the winding-based discharge methods becomes complicated because not all of them are applicable to any EV powertrains when the system parameters are considered, which has been somewhat illustrated in Chaps. 3 and 4. Consequently, a new issue that how to select an appropriate winding-based discharge method (especially in the running situations) for a new EV according to the system parameters arises.

This chapter proposes general rules to pre-evaluate whether or which winding-based discharge technique can be adopted for a particular EV, contributing to discharge method selection over the high-speed range. The proposed criteria are built upon three principles. Firstly, "five-second discharge" is the main requirements to determine whether a discharge approach is appropriate. Secondly, if a winding-based discharge is effective when a crash request occurs at the highest speed (ω_{rated} in this study), it must be effective for all of the running situations. Thus, when developing the theoretical discharge method selection rules, the highest speed situation is considered. Thirdly, the pure winding-based methods are preferred compared to the hybrid one due to the advantages of lower cost and size. It deserves to be mentioned that the proposed selection rules are especially able to judge whether an NDZQ method and an NDNQ method can be adopted in particular EV drives.

6.2 Selection Principles for NDZQ Method

From the perspective of discharge effect, as for the NDZQ methods discussed in Chaps. 3 and 5, after injecting a large *d*-axis current into the machine, the bus voltage instantly decreases under U_{safe}. However, it needs to be mentioned that although "instant discharge" is the originally expected performance for an NDZQ method, there still exists another acceptable "long-cycle discharge" scenario for this method, which can be described as that the bus voltage is higher than 60 V at the moment when discharge starts, but after bleeding for several seconds (within 5 s), it can become no larger than U_{safe}. In this part, the parametric conditions that are able to be used for determining whether the NDZQ discharge techniques can be applied to a particular PMSM drive used in EVs are presented.

6.2.1 Instant Discharge Occasions

Although the NDZQ methods with instant discharge characteristics have been illustrated previously, the relevant explanations are so diffuse that it is difficult for a reader to follow. Thus, comprehensive descriptions of the NDZQ methods are given at first. NDZQ current control contains two modes, namely, traditional LAD-CI and voltage regulation-based methods, of which simplified schematic diagrams can be depicted as Fig. 6.2a, b, respectively. Although both of them are based on flux-weakening principle, the discharge processes are totally different. As for the traditional LAD-CI method, a fixed negative value will be set as the *d*-axis current reference (i_{d_ref}) after a discharge request occurs. For the sake of simplicity, i_{d_ref} is usually set as the lower limit ($-I_{safe}$) of the system safe current in engineering. Figure 6.3a illustrates the desired DC-bus voltage performance when the traditional LAD-CI method works successfully. Immediately when the current is injected into the machine, the bus voltage is expected to experience a sudden drop under 60 V, though the machine speed does not change greatly. The period between t_0 and t_1 represents the current response time (can be regarded as discharge time for this method) which is pretty short, and the instantaneous voltage at t_1 is denoted as balanced voltage (BV). After

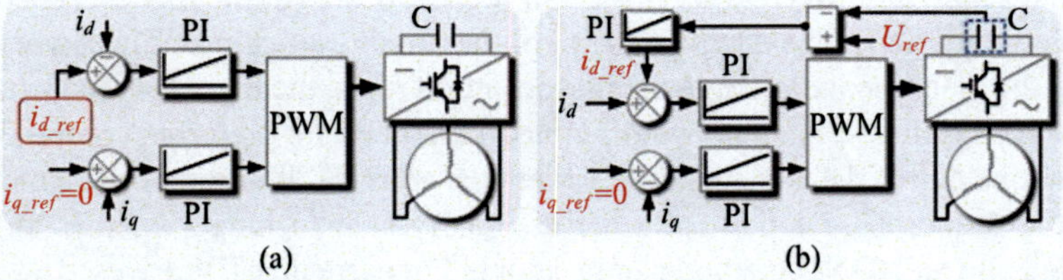

Fig. 6.2 NDNQ methods. a LDA-CI strategy. b Voltage regulation-based strategy

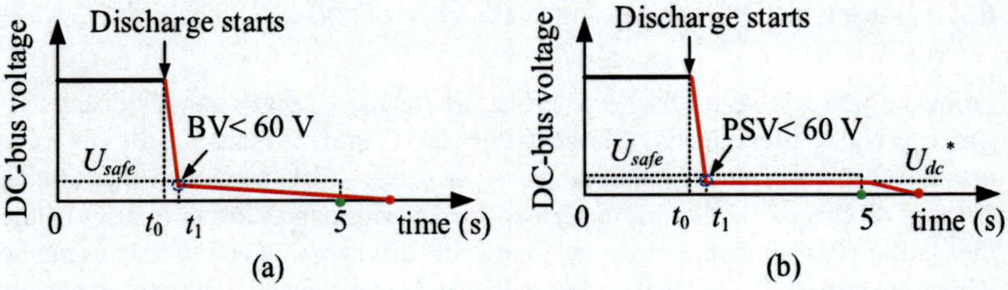

Fig. 6.3 Expected discharge process of instant discharge. a LDA-CI strategy. b Voltage regulation-based strategy

the DC-bus voltage reaches BV, it will constantly decrease in proportion to the declining speed of PMSM. Compared to Fig. 6.2a, the reference d-axis current of the voltage regulation-based method is generated by using a voltage controller (PI). Similarly, two current controllers (PI) are employed for current control. The features of the voltage controller and current controllers are described as follows: (1) The input of the voltage controller is the error between the pre-set voltage (PSV) U_{ref} and the measured value U_{dc}, and the magnitude of the output needs to be restrained within the safe working range (no larger than I_{safe}); (2) In Fig. 6.2a, the main function of the current controllers is to maintain the d, q-axis currents at the desired levels during discharge. In Fig. 6.2b, although the DC-bus voltage is one of the direct control variables, voltage reduction is still achieved by regulating the currents in the machine in virtue of the flux-weakening principle. Therefore, during discharge, the current controllers are used to track the output of the voltage controller which is not a constant and the q-axis reference, respectively; (3) Considering that when designing the internal parameters of the current controllers for the normal operations, both the stability and dynamics of the PMSM drive are usually taken into account, the integral and gain factors used in the discharge process can be consistent with the ones used in the normal control process. Figure 6.3b shows that the DC-bus voltage is expected to track the PSV after the voltage control algorithm is executed, and this will last until the machine speed is too low to generate sufficient back EMF. Hopefully, the current response time is short as well. According to the above analysis, it can be found that the NDNQ methods can totally reduce the DC-bus voltage within five seconds as long as the system parameter I_{safe} is large enough. Specifically, if the allowable injected d-axis current i_{d_ref} (limit value is $-I_{safe}$) is capable of reducing the back EMF of the PMSM under U_{safe} at the highest rotating speed ω_{rated}, the NDZQ current control techniques can be selected to discharge the DC-bus capacitor within a short period for the EV drive during crash.

As for a PMSM whose electrical model is shown in (3.7) and (3.8), when using an NDZQ method to discharge the DC-bus capacitor, assume that the required injected d-axis current is i_{d_ref} and when the system gets stable at the speed of ω_{rated}, the following equations can be derived:

$$\begin{cases} u_d = i_{d\_ref} R_s \\ u_q = p\omega_{rated} L_d i_{d\_ref} + \Psi_f \end{cases} \tag{6.1}$$

Moreover, according to Fig. 6.3 and (3.10), the DC-bus voltage U_{dc} should also meet the following condition during discharge:

$$U_{dc} = \sqrt{u_d^2 + u_q^2} \le U_{safe} \tag{6.2}$$

Substitute (6.1) into (6.2), and the available solution of the in equation is:

$$i_{d\_ref} \le f_{sol}(p, L_d, \Psi_f, \omega_{rated}, R_s) < 0 \tag{6.3}$$

where f_{sol} represents the expansion form of the solution, that is,

$$f_{sol} = \frac{-p^2 L_d \Psi_f \omega_{rated}^2 + \sqrt{-p^2 R_s^2 \Psi_f^2 \omega_{rated}^2 + p^2 L_d^2 U_{safe}^2 \omega_{rated}^2 + R_s^2 U_{safe}^2}}{p^2 L_d^2 \omega_{rated}^2 + R_s^2} \tag{6.4}$$

It can be noticed that firstly, the required injected d-axis current is restrained by not only the system safe current but also the other machine parameters (p, L_d, Ψ_f, R_s, ω_{rated}). Secondly, in order to adopt an NDZQ strategy to achieve instant discharge, the system safe current should cover the required range, that is,

$$-I_{safe} \leq f_{sol} \tag{6.5}$$

It deserved to be mentioned that when performing the discharge actions, the optimal range of the d-axis current reference for the LDA-CI method is $[I_{safe}, f_{sol}]$. Moreover, as long as Eq. (6.5) is satisfied, the output of the voltage controller (d-axis current reference) for the voltage regulation-based method is able to reach the position that can make the back EMF of PMSM under U_{safe} instantly.

6.2.2 Long-Cycle Discharge Occasions

As for the PMSM drives that do not meet the condition (6.5), the bus voltage cannot get down under U_{safe} immediately even when the maximum allowable current $-I_{safe}$ is injected. But as the residual energy is continuously consumed, the bus voltage for some systems might reduce to 60 V in 5 s at most, of which voltage performances are shown in Fig. 6.4. In these cases, NDZQ current control algorithms are still available. Theoretically, in Fig. 6.4, the red line (case a) is the critical case whether a NDZQ method can be used for discharge for an EV drive.

When the d, q-axis currents are set as $-I_{safe}$ and 0, respectively, according to (3.4), the energy consumed by the internal windings (Q_{bin}) during the discharge process (within t_r) can be calculated by:

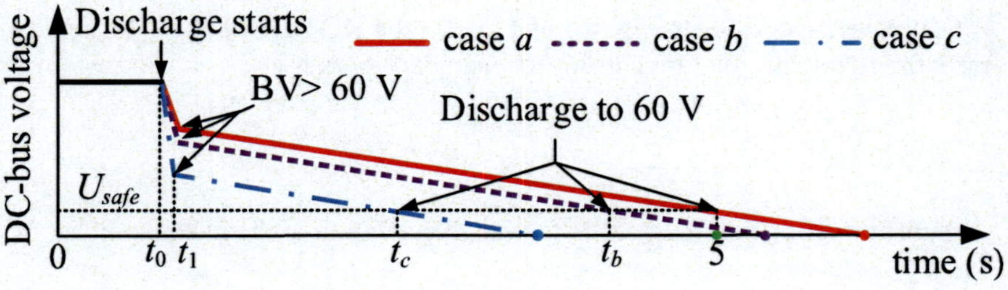

Fig. 6.4 Discharge process of long-cycle discharge

$$Q_{bin} = (i_d^2 + i_q^2)R_s t_r = I_{safe}^2 R_s t_r \tag{6.6}$$

In this study, because t_r is the required discharge period before the DC-bus voltage reaches U_{safe}, it is 5 s at most. In addition to the machine windings, with reference to (3.2), the mechanical friction can also consume energy, which can be described as:

$$Q_{bfri} = \int_{t_r} F \cdot \omega_m^2 \mathrm{d}t \tag{6.7}$$

Assuming that the machine speed declines in a linear trend during discharge, $\omega_m$is written as:

$$\omega_m(t) = \omega_{rated} - \frac{\omega_{rated} - \omega_{m\_th}}{t_r} \cdot t \tag{6.8}$$

ω_{m_th} can be given directly in practice. If it is not given, with reference to (3.10), ω_{m_th} can be estimated by:

$$\omega_{m\_th} = \frac{U_{safe}}{\sqrt{3}C_e'(\Psi_f - L_d I_{safe})} \tag{6.9}$$

It deserves to be mentioned that once the calculated value for ω_{m_th} is over ω_{rated}, let $\omega_{m\_th} = \omega_{rated}$. In this case, (6.5) must be satisfied. Substitute (6.8) into (6.7), and the total energy that can be dissipated by the machine windings and mechanical friction within 5 s can be derived as:

$$Q_{tot} = Q_{bin} + Q_{bfri} = 5I_{safe}^2 R_s + \frac{5F(\omega_{rated}^2 + \omega_{rated}\omega_{m\_th} + \omega_{m\_th}^2)}{3} \tag{6.10}$$

For an EV drive system, with reference to (4.13), when rotating speed drops from the highest value to the threshold speed, the energy that requires to be consumed is:

$$Q_b = \frac{1}{2}J(\omega_{rated}^2 - \omega_{m\_th}^2) + \frac{1}{2}C(U_{dc0}^2 - U_{safe}^2) \tag{6.11}$$

Definitely, as long as the system discharge capacity (when $-I_{safe}$ is set as the d-axis current reference) exceeds the demand, the NDZQ techniques can be used to discharge the DC-bus capacitor in a long-cycle period. Thus, the criterion for discharge technique selection is:

$$Q_b \le k_1 Q_{tot} \tag{6.12}$$

where k_1 is a reliability coefficient and it equals 1 theoretically. But considering that as the bus voltage decreases, the total current in the machine cannot always remain at I_{safe}, leading to relatively lower discharge capacity. Hence, k_1 can be set as less than 1 in engineering.

6.2.3 Implementation Procedures of Selection Rules for NDZQ Methods

To determine whether a NDZQ method can be selected to discharge the DC-bus capacitor of a particular PMSM drive system by using (6.5) and (6.12), five steps need to be implemented (see Fig. 6.5):

(1) Use the machine parameters p, L_d, ψ_f, R_s and ω_{rated} to compute the required d-axis current threshold f_{sol}.
(2) Compare f_{sol} with the system safe current. If the system can withstand the current with a magnitude of f_{sol}, it can be concluded that the NDZQ method is suitable for the instant discharge. Otherwise, execute step 3).
(3) Substitute the values of U_{safe}, Ψ_f, L_d, I_{safe} and C'_e into (6.9) to calculate the threshold speed.
(4) Calculate the total energy Q_{tot} that can be consumed using (6.10) and employ (6.11) to obtain the energy Q_b that needs to be dissipated.

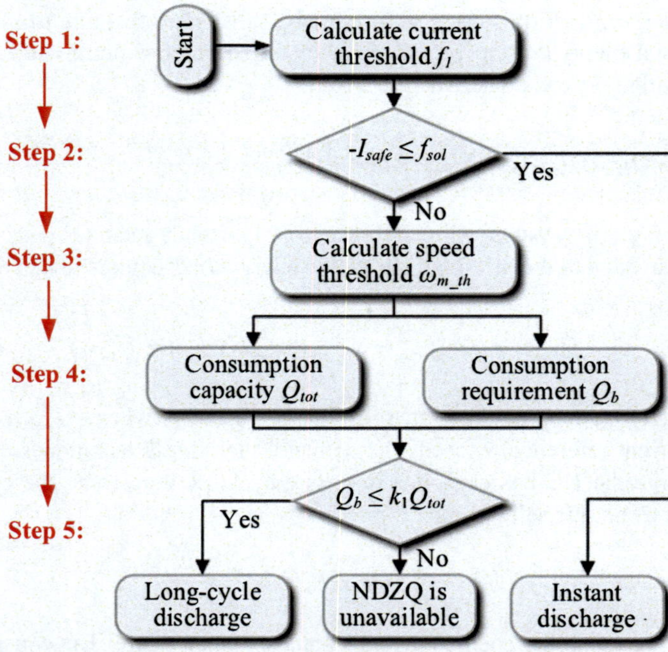

Fig. 6.5 Implementation procedures of NDZQ method selection

(5) Compare Q_{tot} with Q_b. If the system discharge capacity is over the discharge burden, the NDZQ method can be used for the long-cycle discharge in theory.

6.3 Selection Principles for Piecewise NDNQ Method

6.3.1 Criteria for Piecewise NDNQ Method Selection

There only exists one effective NDNQ discharge method so far, that is, piecewise NDNQ method. The rationale behind it is that by injecting a relatively large negative q-axis current, a high braking torque will be generated in the PMSM, leading to a rising deceleration. As long as the machine speed can arrive at the threshold ω_{m_th} before 5 s, the discharge requirement is met. The NDNQ method focuses on controlling the q-axis current rather than the d-axis current (like NDZQ methods), and it is definitely a long-cycle discharge strategy.

As is illustrated in Chap. 3, assume that the period of each segment is Δt, and the required negative q-axis current $i_{q\_ref\_i}$ for each segment is shown in (3.19). When the system is under control, according to (3.12) and (3.13), the electromagnetic torque T_{e_i} and deceleration of the PMSM in each segment can be described as:

$$T_{e\_i} = 1.5p\Psi_f i_{q\_ref\_i} \tag{6.13}$$

$$a_{dec\_i} = \frac{T_{e\_i}}{J} \tag{6.14}$$

Further, the real-time machine speed during discharge is expressed as:

$$\begin{cases} \omega_m(t) = \underbrace{\omega_{rated} + \sum_{i=1}^{n+1} a_{dec\_i-1}\Delta t}_{\omega_{mi-1}} + a_{dec\_n+1}(t - n \cdot \Delta t) \\ n = Quotient\left(\frac{t}{\Delta t}\right) \end{cases} \tag{6.15}$$

where $a_{dec\_0} = 0$. The expression of the initial speed ω_{mi-1} at the start of each segment can be also seen in (6.15).

For the purpose of "five-second discharge", the machine speed at 5 s should be less than the threshold. Hence, the following inequality condition needs to be satisfied:

$$\begin{cases} \omega_{m\_tr} = \omega_{rated} + \sum_{i=1}^{n_{tr}+1} a_{dec\_i-1}\Delta t + a_{dec\_n_{tr}+1}(5 - n_{tr} \cdot \Delta t) \le \omega_{m\_th} \\ n_{tr} = Quotient\left(\frac{5}{\Delta t}\right) \end{cases} \tag{6.16}$$

It deserves to be mentioned here that ω_{m_tr} is the estimated machine speed at 5 s. ω_{m_th} is the threshold speed that can be calculated by (6.9) with I_{safe} replaced by $i_{d\_ref\_ntr+1}$ [obtained by using (3.20)] or given directly.

Equation (6.16) can be regarded as the criteria for judging whether a piecewise NDNQ discharge technique can be applied to a PMSM drive. Clearly, the number of pole pairs, flux linkage of PM, the maximum rotating speed, system safe current, phase resistance, rotor inertia and the pre-set period of each segment are closely related to the applications of an NDNQ method. Besides, it is interesting that the initial DC-bus voltage has little to do with this strategy.

6.3.2 Implementation Procedures

When using the proposed criteria to determine whether a NDNQ strategy can be adopted for discharging the DC-bus capacitor of a PMSM drive in post-crash conditions, the implementation process is divided into the following three steps (see Fig. 6.6):

(1) Preset the period of each segment Δt (e.g., 0.5 s) and predict the required q, d-axis current references $i_{q\_ref\_i}$, $i_{d\_ref\_i}$ using (3.19) and (3.20).
(2) Use the machine parameters J, p, Ψ_f, U_{safe}, L_d, C'_e and ω_{rated} to compute the estimated speed ω_{m_tr} at 5 s and the threshold speed ω_{m_th}.
(3) Compare ω_{m_tr} and ω_{m_th}. If the estimated speed is lower than the threshold, it is concluded that the NDNQ method is suitable for discharge. Otherwise, it is not applicable.

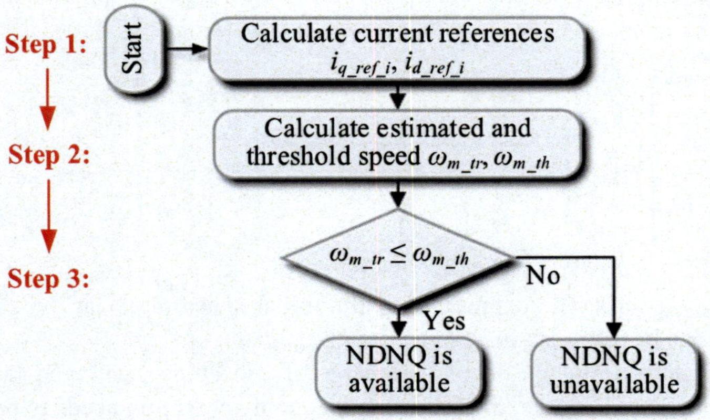

Fig. 6.6 Implementation procedures of NDNQ method selection

6.3.3 Overall Discharge Technique Selection Rules

Equations (6.5), (6.12) and (6.16) just illustrate the analytical applicable occasions of different pure-winding-based discharge techniques in the high-speed situations. But in engineering, more aspects need to be taken into account when using these criteria to select one effective discharge way for a particular PMSM drive. Firstly, compared to the piecewise NDNQ method that requires many calculations, the NDZQ methods are more preferable due to its simplicity. Secondly, although the long-cycle NDZQ approach and the piecewise NDNQ method target at different variables (d-axis and q-axis currents, respectively), the windings can consume the same amount of energy within 5 s because the total injected current is expected to equal I_{safe} regardless of either scheme. Consequently, if (6.12) is satisfied, there must exist a set of q-axis current references [calculated by (3.19)] that satisfy (6.16) for a PMSM in theory, and vice versa. In other words, the maximum discharge capacity of the two long-cycle discharge methods is identical in theory. Another issue that how to determine which one is better arises. Chapter 3 has proven that due to the working properties of a PMSM, as the DC-bus voltage decreases, it is difficult for an NDZQ method to maintain the d-axis current at the maximum level (the total current gradually becomes lower), while the NDNQ method can achieve longer large-current discharge process. Therefore, in the actual applications, the NDNQ strategy is inclined to show higher discharge capacity in reality. According to the above analysis, the aforementioned problem concerning long-cycle discharge strategy selection can be solved by properly setting the value of k_1 (artificially reducing the discharge capacity of the NDZQ method). Because it is unnecessary to accurately design k_1, an empirical value of 0.65 is set in this study.

Overall, the discharge technique selection rules can be summarized as the following sequential procedures (see Fig. 6.7): (1) First, if (6.5) is satisfied, an instant NDZQ discharge algorithm is adopted due to its fast voltage reduction speed regardless of the states of (6.12) and (6.16). (2) Then, if (6.12) is satisfied when $k_1 = 0.65$, the long-cycle NDZQ discharge method is used without considering whether (6.16) is met because the procedures of piecewise NDNQ implementation are much more complicated. (3) If (6.12) is unsatisfied when $k_1 = 0.65$, while (6.16) is satisfied, the piecewise NDNQ method will be employed. (4) Once the parameters of the EV powertrain do not meet all the derived requirements, the hybrid discharge method is supposed to be employed for DC-bus capacitor discharge because it is parameter-independent.

6.4 Case Studies and Results

Different EV propulsion systems have different parameters. In order to validate the proposed criteria and discharge technique selection rules as sufficiently as possible, in this study, case studies (six cases) are employed for verifications. The verifications

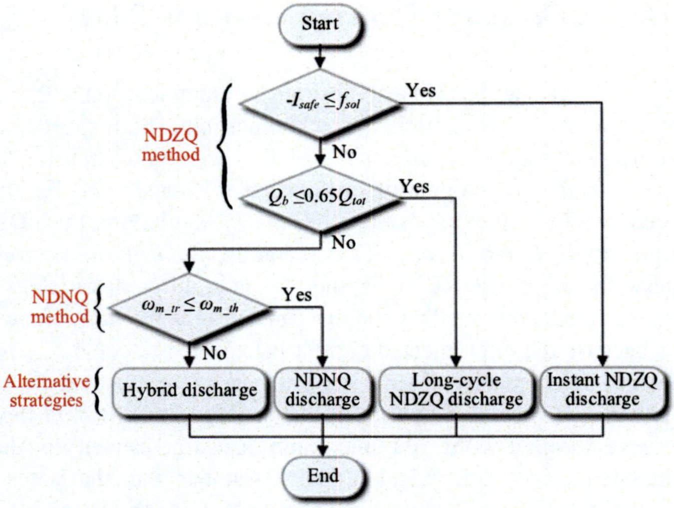

Fig. 6.7 Overall discharge technique selection procedures

can be grouped into two categories. Firstly, the overall selection rules are adopted to pick out appropriate winding-based discharge techniques for four different PMSM drives (case 1–case 4) which have been studied in [152–154], and the selected discharge scheme for each system is tested by simulation. Secondly, because the winding-based discharge methods used in Chaps. 3 and 4 have been proven to be effective, as long as it can be further proven that the system parameters in these chapters comply with the criteria for selecting the corresponding winding-based discharge strategies, the proposed criteria can be considered to be effective to some degree. Therefore, the proposed criteria are used to judge whether the selected winding-based discharge strategies for the systems in Chaps. 3 and 4 (case 5 and case 6), which took parameter influence into account, are suitable, and these can be treated as the experimental verifications in this chapter.

6.4.1 Verifications of Winding-Based Discharge Method Selection Rules

The parameters of the four new cases are shown in Table 6.2.

(a) **Case 1**

According the procedures in Figs. 6.5 and 6.7, substitute the parameter values into (6.4) and it can be obtained that $f_{sol} = -18.2$ A. Because the magnitude of the required d-axis current is smaller than the safe current (criterion (6.5) is satisfied), the bus voltage would sharply decrease down to 60 V after injecting the required

Table 6.2 PMSM powertrain parameters of case 1–4

Parameters	Case 1	Case 2	Case 3	Case 4
Stator winding resistance R_s (Ω)	2.4	0.021	0.29	0.011
Number of pole pairs p	4	4	5	4
Moment of inertia J (kg·m^2)	0.065	0.35	0.016	0.165
Viscous coefficient F	0.0035	0.004	0.0015	0.0035
Permanent magnet flux linkage Ψ_f (Wb)	0.098	0.236	0.24	0.075
DC-bus voltage U_{dc0} (V)	400	500	540	334
System safe current I_{safe} (A)	20	300	15	150
Rated speed ω_{rated} (rad/s)	523	220	270	430
DC-bus capacitor$C$ (µF)	560	4700	1000	2400

d-axis current in theory. When a discharge request occurs at the maximum speed, the discharge performances of the LDA-CI ($i_{d\_ref} = f_{sol}$) and the voltage regulation-based methods are shown in Fig. 6.8a, b, respectively. Importantly, both NDZQ methods are able to reduce the bus voltage under 60 V immediately when the discharge algorithms are executed, and the current response time is nearly 40 ms. This proves that the criterion (6.5) is effective. Before leaving Fig. 6.8, it can be seen that there are obvious differences between the two NDZQ discharge strategies. Figure 6.8a illustrates that the bus voltage continuously decreases after injecting the fixed *d*-axis current, and it will reach zero at about 4.4 s, after which the bus voltage fluctuates around zero. At 5.5 s, the rotating speed is about 250 rad/s. In terms of the voltage regulation-based algorithm, it can maintain the DC-bus voltage at the pre-set value (56 V in this study) until 5.5 s, but the final machine speed is 350 rad/s. These represent that more energy has been dissipated by the machine windings when using the LDA-CI method in the discharge process. Consequently, the energy bleeding rate of the LDA-CI method is larger than that of the voltage regulation-based strategy. The reasons why this phenomenon arises can be explained by the current characteristics. In detail, when the voltage regulation-based algorithm is launched, the *d*-axis current quickly becomes −18.3 A, but afterwards, its magnitude gradually decreases because as the speed decreases, the flux-weakening current magnitude should get down simultaneously according to the PMSM intrinsic properties so as to generate the same back EMF (maintain at the pre-set bus voltage). Comparatively speaking, the *d*-axis current in Fig. 6.8a stabilizes at −18.2 A until about 4.4 s. Simultaneously, although the *q*-axis current reference is set as zero, it automatically turns to be negative due to the cross-coupling effect during discharge. In Fig. 6.8a, the *q*-axis current in the machine is about −3.2 A at 0.5 s, but its magnitude increases constantly. However, the *q*-axis current in Fig. 6.8b keeps −3.2 A. In summary, the current of the LDA-CI method is larger than that of the voltage regulation-based method due to the machine intrinsic properties so that it consumes more energy. Since instant discharge can be achieved for Case 1, the instant NDZQ discharge method will be selected for discharging the system according to the overall selection rules in Sect. 6.3.3.

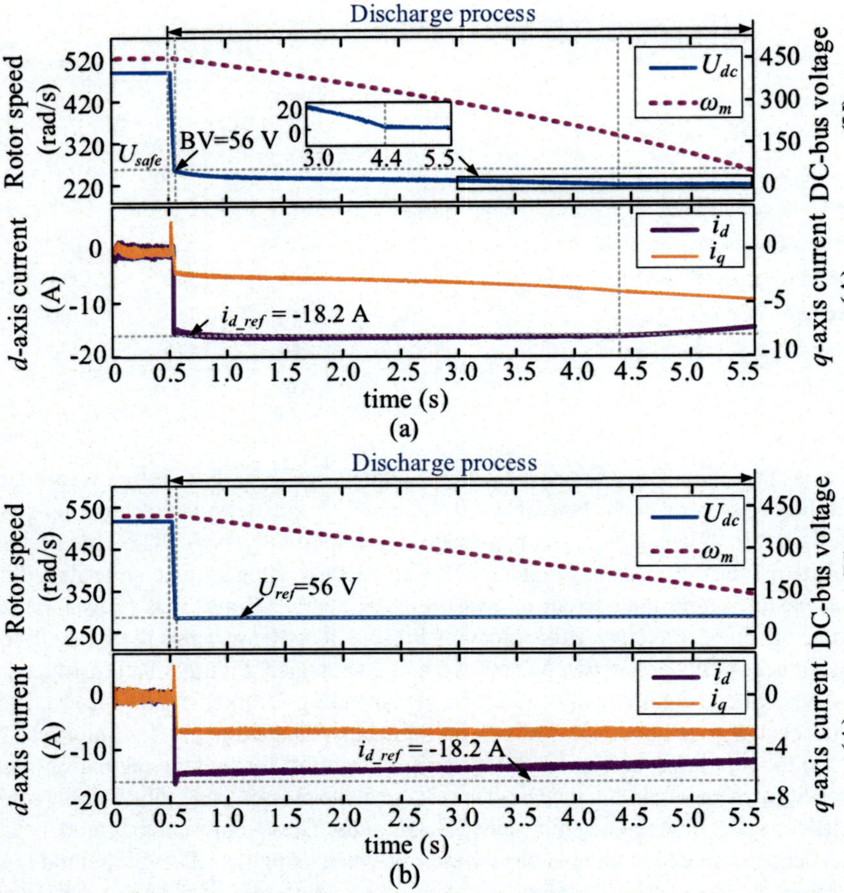

Fig. 6.8 Discharge performances of Case 1 when NDZQ methods are used. **a** LDA-CI method. **b** Voltage regulation-based method

(b) Case 2

Substitute the system parameters into (6.4), and $f_{sol} = -343$ A. Obviously, I_{safe} does not cover the required range, so instant discharge cannot be achieved for Case 2. Then, $k_1 Q_{tot}$ and Q_b can be calculated according to (6.10) and (6.11) as 9396 and 9049 J, respectively. Because the criterion (6.12) is satisfied, the long-cycle NDZQ discharge process should be employed. Figure 6.9 shows the discharge characteristics of the LDA-CI ($i_{d\_ref} = -I_{safe}$) and the voltage regulation-based methods. First of all, the bus voltage only reduces to 120 V when −300 A *d*-axis current is injected, proving that the instant discharge is not available. Secondly, the capacitor voltage decreases to 60 V at about 2.45 s where the rotating speed is 125 rad/s in Fig. 6.9a, and the bus voltage reaches the reference level at nearly 2.55 s for the voltage-regulation strategy. These indicate that the proposed discharge selection rules have selected the proper

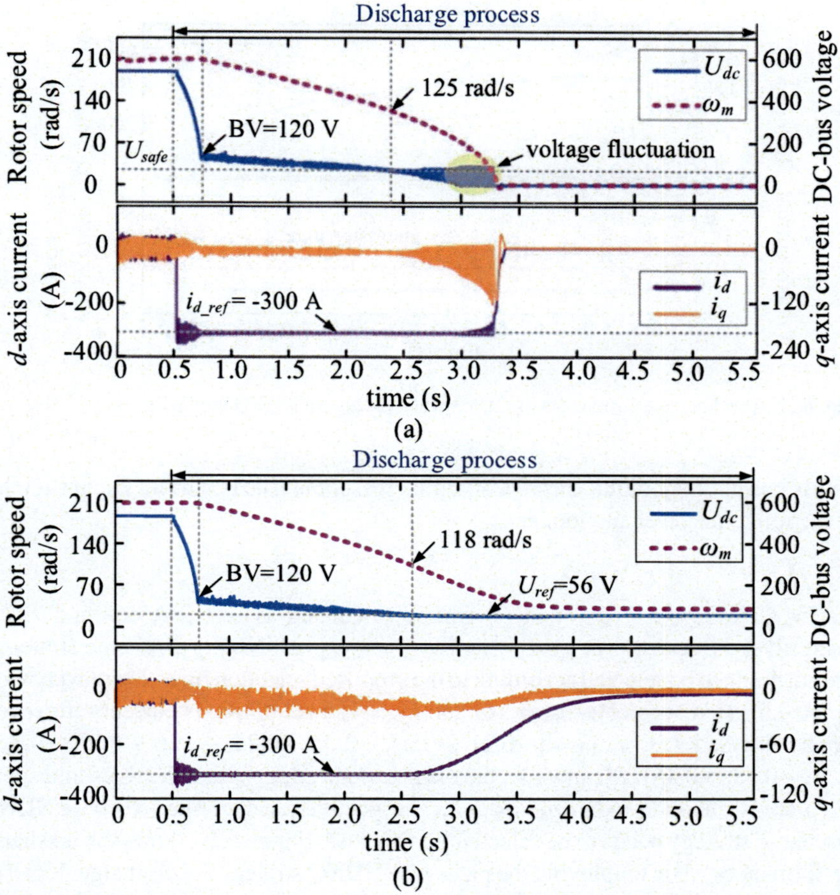

Fig. 6.9 Discharge performances of Case 2 when NDZQ method is used. **a** LDA-CI method. **b** Voltage regulation-based method

bleeding algorithm for this PMSM drive. Then, similar as the results in Fig. 6.8, the LDA-CI method presents faster energy bleeding speed. For Case 2, at about 3.4 s, the residual energy can be completely consumed in Fig. 6.9a. Finally, between 3.0 s and 3.4 s, the magnitude of the real d-axis current in the machine is decreasing. Although the q-axis current witnesses a slight increase, the total current in the machine still drops, revealing the necessity of the reliability coefficient k_1. Finally, it is interesting to see that the q-axis current and the bus voltage fluctuate in Fig. 6.9a at around 3.0 s. This happens because the bus voltage has arrived at the relatively low position at the moment, which is hard to maintain the d-axis current at the high reference level. In this case, the q-axis current magnitude will automatically increase (more energy flows to the bus side) to ensure that the DC-link voltage is as high enough as possible to stabilize the d-axis current. Whereas, when the bus voltage is very low, even the

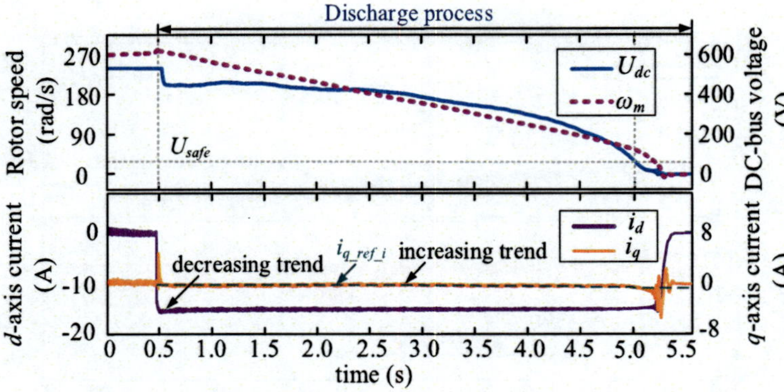

Fig. 6.10 Discharge performances of Case 3 when piecewise NDNQ method is used

q-axis current magnitude shows a sharp increase, the d-axis current cannot remain at the reference level any longer.

(c) **Case 3**

As for Case 3, f_{sol}, k_1Q_{tot} and Q_b can be calculated as -16.2 A 468 and 727 J, respectively. Because $-f_{sol} > I_{safe}$ and $k_1Q_{tot} < Q_b$, the NDZQ discharge strategies are not going to be selected according to the proposed selection rules. Moreover, when $\Delta t = 0.5$ s (ten segments, $n_{tr} = 10$) and $k_2 = 1$, a set of q-axis current reference s can be obtained: $i_{q\_ref\_1...10} = \{-0.21, -0.22, -0.23, -0.24, -0.26, -0.27, -0.3, -0.33, -0.38, -0.45\}$. Meanwhile, the estimated speed ω_{m_tr} and the threshold speed ω_{m_th} are 108 and 119 rad/s ($\omega_{m\_tr} < \omega_{m\_th}$), respectively. Hence, the piecewise NDNQ discharge strategy needs to be selected for this case. Figure 6.10 shows the discharge performances when employing the piecewise NDNQ strategy for discharge. It can be seen that the bus voltage gets down to 60 V at about 5.5 s and no dramatical voltage surge is witnessed (fluctuations are caused by the impact of Δt). On these grounds the proposed discharge technique selection rules pick out the proper energy bleeding method for Case 3.

(d) **Case 4**

When the PMSM system parameters of Case 4 are used for calculation, it can be obtained that $f_{sol} = -201$ A ($-f_{sol} > I_{safe}$); $k_1Q_{tot} = 1907$ J and $Q_b = 15{,}384$ J ($k_1Q_{tot} << Q_b$); when $\Delta t = 0.5$ s, $i_{q\_ref\_1...10} = \{-1.92, -1.94, -1.95, -1.96, -1.97, -1.99, -2.0, -2.01, -2.03, -2.04\}$, $\omega_{m\_tr} = 403$ rad/s and $\omega_{m\_th} = 138$ rad/s ($\omega_{m\_tr} > \omega_{m\_th}$). Noted that none of the criteria are satisfied, so it will be concluded that the pure-winding-based discharge techniques cannot be applied to the drive according to the proposed discharge method selection rules (hybrid method is preferred). To verify this conclusion, Fig. 6.11 gives the simulation results when the instant NDZQ discharge (LDA-CI) method is adopted after a discharge request arises. It can be seen that after injecting current, the DC-bus voltage gradually increases because much more mechanical energy is converted to the electrical energy, recharging the

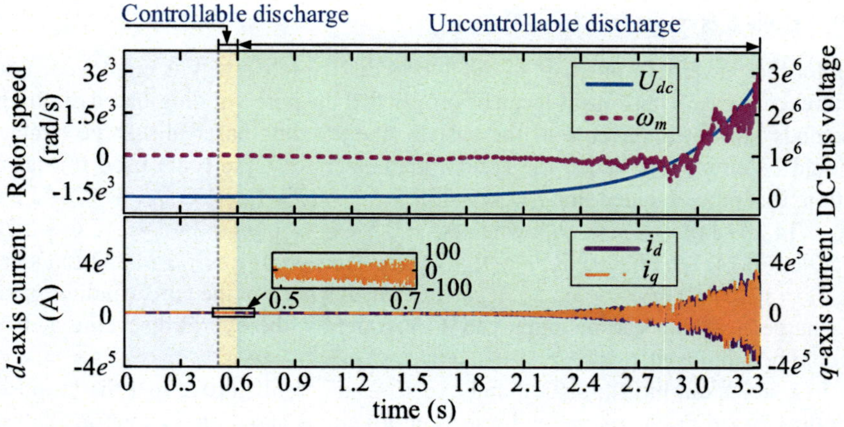

Fig. 6.11 Discharge performances of Case 4 when LDA-CI method is used

DC-bus capacitor. This happens because although only a negative *d*-axis current is injected, a relatively large *q*-axis current will be automatically generated due to the cross-coupling effects during discharge. The *q*-axis current is able to produce large braking torque, quickening the speed deceleration process. At the moment, the discharge capacity of the windings is much lower than the conversion rate when only a total of I_{safe} is applied to the machine. Finally, as the bus voltage rises, the current in the machine is inclined to become uncontrollable, failing to reach the DC-bus capacitor discharge goal. All in all, the pure-winding-based discharge method is not applicable to Case 4, and the hybrid or bleeder-based techniques should be utilized, complying with the proposed selection rules.

6.4.2 Judgement for Discharge Methods in Previous Chapters

(a) **Case 5 (system in Chap. 3).**

As for the powertrain in Chap. 3, the piecewise NDNQ discharge strategy is employed. Here, using the parameters of the powertrain in Chap. 3, the proposed criteria for selecting the piecewise NDNQ method are verified. Following the procedures in Fig. 6.6, when $\Delta t = 0.5$ s, the ten *q*-axis current references are calculated as $i_{q\_ref\_1\ldots10} = \{-10.6, -11.3, -12.1, -13.1, -14.4, -16.2, -18, -23.4, -35.4, -39.5\}$, and $i_{d\_ref\_1\ldots10} = \{-99.4, -99.3, -99.2, -99.1, -98.9, -98.6, -98.3, -97.2, -93.5, -91.8\}$. Then, it can be calculated that $\omega_{m\_tr} = 66.6$ rad/s and $\omega_{m\_th} = 114$ rad/s ($\omega_{m\_tr} > \omega_{m\_th}$). Hence, it can be concluded that the NDNQ method is applicable to the system in Chap. 3, so the criteria for piecewise NDNQ method are effective.

(b) **Case 6 (system in Chap. 4)**

As for the powertrain in Chap. 4, the hybrid discharge strategy is employed. With reference to case 4, as long as it can be proven that the pure-winding-based discharge methods are not applicable to the system, the winding-based discharge strategy should be employed. When the system parameters of Case 6 are used for calculation, it can be obtained that $f_{sol} = -158.5$ A ($-f_{sol} > I_{safe}$); $k_1 Q_{tot} = 7764$ J and $Q_b = 14{,}309$ J ($k_1 Q_{tot} << Q_b$); when $\Delta t = 0.5$ s, $i_{q\_ref\_1 \ldots 10} = \{-5.4, -5.6, -5.76, -5.94, -6.14, -6.36, -6.61, -6.9, -7.2, -7.58, -8.01\}$, $\omega_{m\_tr} = 237$ rad/s and $\omega_{m\_th} = 120$ rad/s ($\omega_{m\_tr} > \omega_{m\_th}$). It can be seen that none of the pure-winding-based discharge strategies can be employed. Consequently, the hybrid discharge method should be employed.

In conclusion, the simulation and experimental results prove that the proposed winding-based discharge method selection principles based on parametric analysis are effective.

6.5 Summary

This chapter studies the principles which are used to select an appropriate winding-based discharge strategy for EVs according to the parameters of their powertrains. Firstly, the winding-based discharge strategies are recategorized. Then, as for the NDZQ methods and NDNQ method, which can be used in the high-speed situations, specific criteria for judging whether they are available for various EVs are derived based on parametric analysis. Then, by synthesizing those criteria, general selection rules for winding-based discharge strategies are developed with engineering needs considered. It deserves to be mentioned that the implementation procedures of the criteria and selection rules have been detailed, through which an engineer can easily design a reliable and cost-effective discharge system for a new EV. The proposed selection rules in this chapter are able to facilitate the application and development of the winding-based discharge techniques, contributing to high crash safety level of EVs.

Chapter 7
Conclusions and Future Work

7.1 Conclusions

As an eco-friendly transportation method, EVs have been witnessing an exponential growth. However, it must be acknowledged that the technologies concerning EVs are not entirely mature and there are still many problems that need in-depth explorations. Notably, the crash safety of EVs is attracting significant attention from not only the scholars and manufacturers but also the ordinary people now, making it a valuable topic for research. In this study, the crash safety-related issues of the high-voltage powertrain-based EVs are addressed. Especially, this research mainly focuses on the high-voltage electric shock hazards arising from the residual energy stored in the power electronic devices and the relevant prevention solutions. In order to dissipate the residual energy, various winding-based DC-bus capacitor discharge strategies are developed for different EVs and crash conditions, which can be regarded as the core technologies and novelties of this study. By using the proposed winding-based discharge methods, the residual energy stored in the EV powertrains can be dissipated quickly (within 5 s) and reliably in a compact and cost-effective way, contributing to improving the crash safety of EVs. The main work and contributions of this study can be summarized as follows:

In Chap. 2, an in-depth literature review on the safety-related topics of EVs in the crash conditions is presented. The review work focuses on the possible injury hazards that the occupants are subjected to and the efforts made to reduce risks. In detail, the injury hazards in the crash conditions are categorized into five groups: physical injuries, electric shock, corrosion, intoxication and burn. The rationale behind these dangers is analyzed firstly. In addition, the international and national regulations that are enacted to deal with these hazards so as to ensure that the vehicle and occupants are as safe as possible during crash are reviewed. It deserves to be mentioned that these regulations are mainly developed to standardize the design, manufacturing and testing processes of the automakers. Finally, the up-to-date practical technologies that can be adopted to reduce the corresponding risks and severity of injuries to passengers

C. Gong, *Crash Safety of High-Voltage Powertrain Based Electric Vehicles*,
Springer Theses, https://doi.org/10.1007/978-3-030-88979-1_7

in the crash conditions are presented. Through the review work, a valuable topic about winding-based discharge techniques is discovered, and further, it is studied in this research. This indicates that the review work can provide guidelines for the future studies. Moreover, it becomes easier for the ordinary people to clearly understand the safety issues of the EVs during crash and what the professionals have done to improve the safety level.

In Chap. 3, a novel piecewise NDNQ discharge method which has stronger discharge capacity practically than the traditional LDA-CI method is developed for the EV powertrains with slightly large inertia and relatively small safe current. In this chapter, in order to illustrate the mechanism of a winding-based discharge strategy, an EFM which is a combination of energetics and electromagnetics is proposed. Through the EFM, the defects of the traditional discharge methods are analyzed by means of case study. Meanwhile, it is found that it is possible to quicken the discharge process by using a NDNQ control algorithm. On this ground a piecewise NDNQ strategy is proposed to quickly discharge the DC-bus capacitor and safely reduce the motor speed without voltage surge. In this part, it can be concluded that firstly, the winding-based discharge method is parameter-dependent. Secondly, if an inevitable crash occurs, the proposed NDNQ approach can avoid high-voltage electric shock hazard for an EV of which powertrain has extreme parameters.

In Chap. 4, a novel combined residual energy consumption strategy based on both the internal windings and the external bleeder circuits is developed for the EV powertrains with highly extreme parameters to achieve five-second discharge requirement without greatly sacrificing the BR size and weight. In this chapter, in order to explicitly evaluate the BR size and weight sacrifice of the traditional external bleeder-based discharge strategy, an accurate model is established firstly. By the use of case study, it is found that for the EV powertrains with highly extreme parameters, the weight and size of the required BR are pretty large, which means that it is necessary to develop a hybrid discharge technique that synthesizes the advantages of the traditional winding-based and bleeder-based discharge schemes. Then, for the hybrid strategy, systematic design procedures are provided, which include external BR design and control algorithm design. Specifically, the external BR resistance is designed based on the extreme working condition (maximum speed), and full-power, partial-power and bleeder-based discharge algorithms are designed according to the working states at the moment of collision. By using the hybrid discharge approach, the size and weight of the external BR can be greatly reduced.

In Chap. 5, reliable winding-based DC-bus capacitor discharge methods which incorporate the PL-FT control strategies are proposed. To achieve a successful discharge over high-speed range without using position sensor, an improved SOSM based on continuous sigmoid function, which has strong robustness against external disturbances, is presented. Moreover, an adaptive sine-wave-based position observer with high immunity to voltage change is proposed for the discharge algorithms in the low- and zero-speed cases. In terms of the discharge procedures, novel indirect current regulation-based discharge methods (voltage regulation-based and speed regulation-based methods) are designed, which has enriched the winding-based discharge theories. Through this research, when implementing a winding-based discharge strategy, the position information measured by the external sensors is not needed any longer. Instead, the position used for achieving a winding-based discharge method can be estimated by the position observers. Because the position loss fault does not influence the discharge process, the reliability of the discharge system is improved. Further, the electric shock injury hazards can be reduced, improving the crash safety of EVs.

In Chap. 6, general principles to determine whether or which winding-based DC-bus capacitor discharge technique is applicable to a particular EV. Because the proposed rules are obtained on the basis of theoretical analysis, they are simple to be implemented in engineering. In detail, firstly, according to the discharge characteristics, the NDZQ method is further divided into the instant and long-cycle NDZQ discharge strategies. By analyzing the desired d-axis current reference of the instant NDZQ method and the discharge capacity of the long-cycle method, the applicable conditions for them are deduced. Then, as for the piecewise NDNQ method, by using the calculated d, q-axis currents which are to be injected into the machine, the real machine speed and threshold speed after five-second discharge can be obtained, which are employed for determining the applicable conditions of this method. Finally, taking the hybrid discharge method into account, the overall rules for determining whether a winding-based discharge strategy is available for a PMSM powertrain are discussed, directly contributing to discharge strategy selection in practice. Undoubtedly, the proposed selection rules in this chapter are important for the promotions of the winding-based discharge strategies.

All in all, in terms of the topic about winding-based DC-bus capacitor discharge, this research combines both theoretical analysis and simulation/experimental validation to investigate and develop comprehensive winding-based DC-bus capacitor discharge theories and techniques for the high-voltage PMSM powertrain-based EVs in the crash conditions. In this area, the following achievements have been made in this study:

- The mechanism of the winding-based discharge strategies is clearly explained.
- In addition to the traditional LDA-CI method, several novel winding-based discharge methods have been developed for different EVs.
- The PL-FT algorithms are integrated into the discharge process, improving the reliability of the discharge system of an EV.
- The impacts of system parameters on the winding-based discharge strategies are explored, based on which the principles for selecting proper discharge strategies for various EV drives are deduced.

7.2 Future Work

Based on the conclusions above and considering the limitations of the existed work, future research could be carried out in the following areas.

- **Further investigation is needed for more crash safety-related problems**

Chapter 2 takes the possible injury hazards to occupants as the main clue, analysing the crash safety-related issues of EVs and the efforts made to prevent injuries. In other words, the safety of occupants is focused on in this study. However, apart from the passengers, the components (especially the fragile electric devices) also need to be protected during crash. Therefore, it is important to study the impacts of crash on the internal devices and the corresponding protection techniques. Moreover, in addition to the electric shock risks which are mainly studied in this research, there are many topics (e.g., RESS-related injuries) that also need to be investigated in the future for the purpose of improving the crash safety comprehensively.

- **More fault-tolerant control algorithms need to be considered for the sake of high reliability**

Chapter 5 integrates the PL-FT control algorithms into the winding-based discharge process. However, during crash, apart from the position loss fault, various failures might occur on the powertrain devices. For instance, the hall-effect current sensors are inclined to malfunction due to the violent impact. In this case, the current sensor fault-tolerant control algorithms need to be employed to ensure a winding-based discharge method can be achieved successfully. Moreover, one or more cables used for connecting the inverter and machine might get broken during crash. At the moment, without effective open-phase fault- tolerant control algorithms, the PMSM cannot be controlled to operate normally. Besides, if an open-phase fault occurs, the discharge capacity of the machine windings will degrade. Even worse, the DC-bus voltage might fail to decrease to the safe level in time. Hence, when different faults are considered during crash, the winding-based discharge strategies need to be optimized by using fault-tolerant techniques.

- **Fully use the switching loss of inverters and skin effect of conductors to accelerate the residual energy consumption and DC-bus voltage reduction velocity**

When using a winding-based discharge strategy, the inverter operates at a high switching frequency (HSF). When the transistors in the inverter switch between "ON" and "OFF" states, switching loss is generated. Therefore, the inverter could consume part of the residual energy when implementing a winding-based discharge method. Besides, it deserves to be mentioned that the influence of HSF goes far beyond this. Specifically, because HSF could generate high-frequency AC harmonics in the phase currents, skin effect occurs when the currents flow through the conductors. In this case, the resistance of the system rises, leading to faster energy consumption rate when the injected currents are fixed. For the sake of preciseness, it is significant to study the relationship between the switching frequency and the switching loss and skin effect. Moreover, the influence of switching loss and skin effect on the winding-based discharge methods needs to be explored.

- **Parameter identification strategies need to be integrated so as to improve the accuracy of the proposed winding-based discharge strategy selection rules**

According to Chap. 6, the winding-based discharge strategy selection rules totally rely on the powertrain parameters, some of which (e.g., the stator inductance, resistance, flux linkage of PM) are variable during not only the discharge process but also the normal control process. Once severe deviations occur on these parameters, it is possible to fail to achieve the "five-second discharge" or "no voltage surge" goals. On this ground it would be better to adopt the real-time parameter identification techniques to obtain the accurate resistance, inductance and flux linkage of the machines during control. Because this research mainly focuses on the basic winding-based discharge selection rules, the parameter uncertainty issue is ignored temporarily.

- **More case studies are needed to verify that the proposed winding-based discharge strategy selection rules are effective**

In Chap. 6, six cases are used to verify that the proposed winding-based discharge strategy selection rules are effective. However, considering that the powertrains are completely different in different EVs, the verifications are not sufficient. In the future, more case studies should be employed to validate the proposed selection rules in engineering.

Appendix

•

1. Vector Control for AC Motors

The basic idea of vector control is coordinate transformation. In detail, the three-phase currents and voltages are transformed to the variables in the two-phase stationary (α, β) reference frame firstly, and then they are transformed to the variables in the direct-quadrature (d, q) rotating reference frame. The classical vector control is achieved by controlling the d, q-axis currents discretely, which are related to the exciting field and torque, respectively. Taking the currents of a PMSM as an example, the main principles concerning vector control are introduced below.

The stator current performance of a PMSM in the three-phase stationary (A, B, C) reference frame is illustrated in Fig. A.1 (black curves), where the coils represent the armature windings. Then, define the two-phase stationary reference frame as the blue curves in Fig. A.1. Assuming that the number of three-phase and two-phase winding turns is N_a and N_b, respectively, it can be obtained that:

$$\begin{cases} N_b i_\alpha = N_a i_A \cos 0 + N_a i_B \cos \frac{2\pi}{3} + N_a i_C \cos \frac{4\pi}{3} \\ N_b i_\beta = N_a i_B \sin \frac{2\pi}{3} + N_a i_C \sin \frac{4\pi}{3} \end{cases} \tag{A.1}$$

In order to obtain a square matrix which is simple for analysis, a virtual current i_0 is added to the two-phase stationary reference frame, that is,

$$i_0 = \frac{N_a}{N_b}(k_{tr} i_A + k_{tr} i_B + k_{tr} i_C) \tag{A.2}$$

where k_{tr} is a constant.

Rewrite (A.1) and (A.2) in the form of matrix, and the relation (Clarke Transformation) between the three-phase currents and two-phase currents is:

C. Gong, *Crash Safety of High-Voltage Powertrain Based Electric Vehicles*, Springer Theses, https://doi.org/10.1007/978-3-030-88979-1

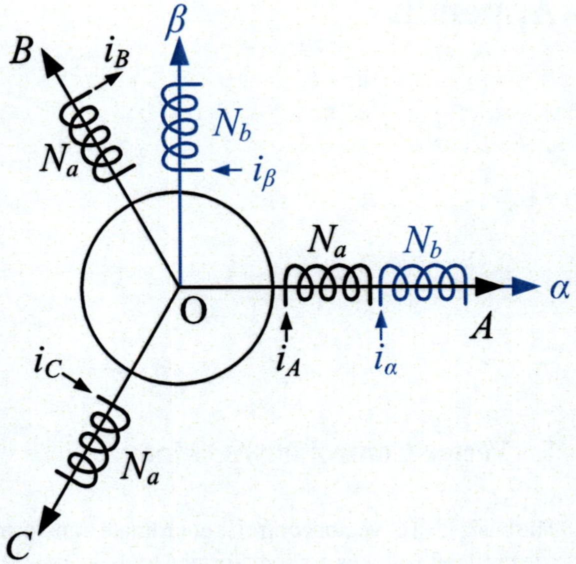

Fig. A.1 Model in stationary reference frames

$$\begin{bmatrix} i_\alpha \\ i_\beta \\ i_0 \end{bmatrix} = \begin{bmatrix} 1 & -\frac{1}{2} & -\frac{1}{2} \\ 0 & \frac{\sqrt{3}}{2} & -\frac{\sqrt{3}}{2} \\ k_{tr} & k_{tr} & k_{tr} \end{bmatrix} \begin{bmatrix} i_A \\ i_B \\ i_C \end{bmatrix} \tag{A.3}$$

Clarke transformation matrix can be extracted from (A.3):

$$C_{clarke} = \frac{N_a}{N_b} \begin{bmatrix} 1 & -\frac{1}{2} & -\frac{1}{2} \\ 0 & \frac{\sqrt{3}}{2} & -\frac{\sqrt{3}}{2} \\ k_{tr} & k_{tr} & k_{tr} \end{bmatrix} \tag{A.4}$$

In the constant power condition, the product of Clarke transformation matrix and its transposed matrix should equal **E**, that is:

$$C_{Clarke} \cdot C_{Clarke}^T = \frac{N_a^2}{N_b^2} \begin{bmatrix} 1 & -\frac{1}{2} & -\frac{1}{2} \\ 0 & \frac{\sqrt{3}}{2} & -\frac{\sqrt{3}}{2} \\ k_{tr} & k_{tr} & k_{tr} \end{bmatrix} \begin{bmatrix} 1 & 0 & k_{tr} \\ -\frac{1}{2} & \frac{\sqrt{3}}{2} & k_{tr} \\ -\frac{1}{2} & -\frac{\sqrt{3}}{2} & k_{tr} \end{bmatrix}$$

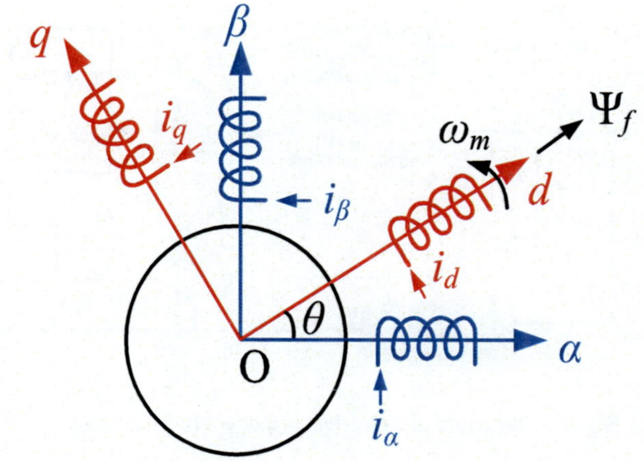

Fig. A.2 Model in two-phase stationary and rotating reference frames

$$= \frac{N_a^2}{N_b^2}\begin{bmatrix} \frac{3}{2} & 0 & 0 \\ 0 & \frac{3}{2} & 0 \\ 0 & 0 & 3k_{tr}^2 \end{bmatrix} = \mathbf{E} \tag{A.5}$$

Therefore, it can be calculated that:

$$\frac{N_a}{N_b} = \sqrt{\frac{2}{3}}, \quad k_{tr} = \frac{1}{\sqrt{2}} \tag{A.6}$$

So far, the three-phase currents have been transformed into the variables in the α, β reference frame. Then, define the direct-quadrature rotating reference frame as the red curves in Fig. A.2. It can be seen that, firstly, the d, q reference frame rotates at the speed of rotor speed, and secondly, the direction of d axis complies with that of the rotor permanent magnet flux linkage. In addition, the angle between the α axis and d axis is the so-called rotor position θ. According to Fig. A.2, the relation (Park Transformation) between the α, β-axis currents and the d, q-axis currents can be described as:

$$\begin{bmatrix} i_d \\ i_q \end{bmatrix} = \begin{bmatrix} \cos\theta & \sin\theta \\ -\sin\theta & \cos\theta \end{bmatrix}\begin{bmatrix} i_\alpha \\ i_\beta \end{bmatrix} \tag{A.7}$$

As for the classical vector control method, after obtaining the variables in the rotating reference frame, the corresponding PMSM model can be established as (3–7), (3–8) and (3–9). Besides, the structure of a classical vector control scheme is shown in Fig. A.3. As for the system, there are several aspects that need to be addressed. Firstly, it can be seen that the d, q-axis currents are controlled discretely. Although a speed controller is used to generate the q-axis current reference in a normal control process, the q-axis current together with the d-axis current can be controlled

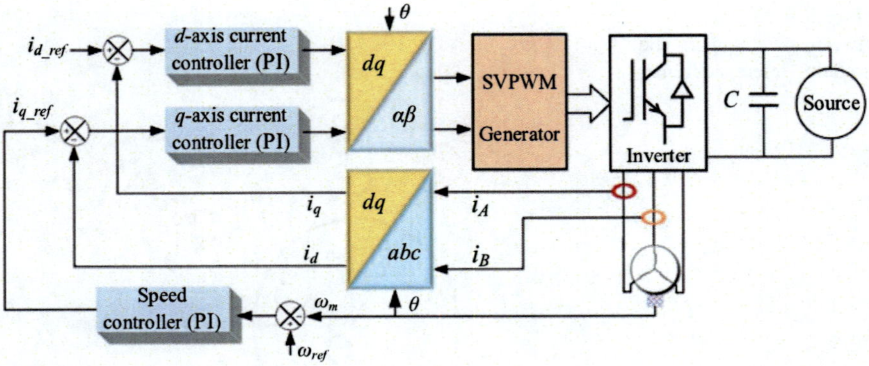

Fig. A.3 Structure of a classical vector control scheme

directly by setting fixed current references when needed. Secondly, the speed and rotor position information are usually measured by sensors which are embedded in the machine. Thirdly, because $i_A + i_B + i_C = 0$ in a machine, it is enough to measure any two of the three phase currents for the coordinate transformation.

2. Awards and Trainings During PhD Period

3. Collaborations During PhD Period

- Participate in the **Newton Advanced Fellowship**, "Artificial Intelligence Supported Safe Operation and Health Prediction for Electric Vehicle Motors", a collaborative project with Prof. G. Wang at School of Electrical Engineering and Automation, Harbin Institute of Technology.

4. Part of Control Codes in TMS320F28335 Platform (Void Main ())

```
void main(void)
{ InitSysCtrl();
  DINT;
  InitPieCtrl();
  IER=0x0000;
  IFR=0x0000;
  delay(50000);
  InitPieVectTable();
  EALLOW;
  SysCtrlRegs.PCLKCR0.bit.TBCLKSYNC=0;
  EDIS;
  EPwm1Regs.ETSEL.bit.INTSEL = ET_CTR_ZERO;
ET_CTR_ZERO=0X1
EPwm1Regs.ETSEL.bit.INTEN = 1;
EPwm1Regs.ETPS.bit.INTPRD = ET_1ST;
EPWMx_INT  ET_CTR_ZERO=0X1
  EPwm1Regs.ETCLR.bit.INT = 1;
  EALLOW;
  PieVectTable.TINT0 = &cpu_timer0_isr;
  PieVectTable.EPWM1_INT = &epwm1_timer_isr;
PieVectTable.SCIRXINTB=&SCIB_RX_ISR;
  EDIS;
  EALLOW;
  GpioCtrlRegs.GPAPUD.bit.GPIO20 = 0;
  GpioCtrlRegs.GPAQSEL2.bit.GPIO20 = 0;
  GpioCtrlRegs.GPAMUX2.bit.GPIO20 = 0;
  GpioCtrlRegs.GPADIR.bit.GPIO20 = 1;
  EDIS;
  InitEPwmGpio();
  Init_Epwm();
  InitCpuTimers();
```

```
        Init_Timer0();
        PieCtrlRegs.PIECTRL.bit.ENPIE=1;
        PieCtrlRegs.PIEIER1.bit.INTx7 = 1;
        PieCtrlRegs.PIEIER3.bit.INTx1 = 1;
        PieCtrlRegs.PIEIER9.bit.INTx3=1;
        IER |=(M_INT3| M_INT1|M_INT9);
        EALLOW;
        SysCtrlRegs.PCLKCR0.bit.TBCLKSYNC=1;
        EDIS;

        InitFlash();
        InitGpio();
        InitXintf();
       InitSci();
        InitSciGpio();
        InitAdc();
        AD2S1210Initiate();
        EnableInterrupts();

        GpioDataRegs.GPBSET.bit.GPIO36 =1;
        GpioDataRegs.GPBCLEAR.bit.GPIO37 =1;
        GpioDataRegs.GPASET.bit.GPIO12 =1;
        GpioDataRegs.GPACLEAR.bit.GPIO13 =1;
        GpioDataRegs.GPASET.bit.GPIO20 =1;

        rampspeed.RampUpMax=500;
        rampspeed.RampDownMax=500;
        ERTM;

       delay(4000);
     while(1)

        {       if(scirx.ucgst==0)
```

```
{
        EPwm4Regs.DBCTL.bit.OUT_MODE=0;
        EPwm4Regs.AQCSFRC.all=0x000A;
}
if(bridge_FLT_constant==0 || scirx.ucgst==1)
{
  EPwm1Regs.DBCTL.bit.OUT_MODE=0;
        EPwm2Regs.DBCTL.bit.OUT_MODE=0;
        EPwm3Regs.DBCTL.bit.OUT_MODE=0;

        EPwm1Regs.AQCSFRC.all=0x000A;
        EPwm2Regs.AQCSFRC.all=0x000A;
        EPwm3Regs.AQCSFRC.all=0x000A;
}
else if (scirx.posset==0)
{
        EPwm1Regs.DBCTL.bit.OUT_MODE=0;
        EPwm2Regs.DBCTL.bit.OUT_MODE=0;
        EPwm3Regs.DBCTL.bit.OUT_MODE=0;
        EPwm3Regs.AQCSFRC.all=0x0000;
        for(pset=0;pset<100;pset++)
{
                EPwm1Regs.AQCSFRC.all=0x000A;
                EPwm2Regs.AQCSFRC.all=0x0006;
                EPwm3Regs.CMPA.half.CMPA=100;
                SET_SPL();
                asm(" RPT #50 || NOP");
                CLR_SPL();
                asm(" RPT #50 || NOP");
           CLR_A0(); CLR_A1();
                asm(" RPT #6 || NOP");
                position0= *EXT_AD2S1210;
        }
```

```
            if(position0>=8000 && position0<=11000)
                {
                        position_cnt=1;
                        position_cnt_s=0;
                        rdc.MotorInitAngle=20154;
                }
            else if(position0>=30000 && position0<=34000)
                {
                        position_cnt=0;
                        position_cnt_s=0;
                        rdc.MotorInitAngle=42884;
                }
            else if(position0>=52000 && position0<=56000)
                {
                        position_cnt=1;
                        position_cnt_s=0;
                        rdc.MotorInitAngle=63822;
                }

            position1=position0;
        }
        else if(scirx.posset==1 && scirx.staset==1 && scirx.controlmodeset==1 &&
scirx.ucgst==0)
        {
            EPwm1Regs.DBCTL.bit.OUT_MODE=DB_FULL_ENABLE;
            EPwm2Regs.DBCTL.bit.OUT_MODE=DB_FULL_ENABLE;
            EPwm3Regs.DBCTL.bit.OUT_MODE=DB_FULL_ENABLE;

            EPwm1Regs.AQCSFRC.all=0x0000;
            EPwm2Regs.AQCSFRC.all=0x0000;
            EPwm3Regs.AQCSFRC.all=0x0000;/
            stopflag=0;
```

```
        if(scirx.dirset==0)
            speed_ref=(float)(scirx.speedset)/BASE_RPM;
        else
            speed_ref=(float)(-1.0)*(scirx.speedset)/BASE_RPM;
    }
    else if(scirx.staset==0)
    {
        speed_ref=10.0;
        if(scirx.controlmodeset==1
        {
            if(rdc.MotorSpeed<=20)
                stopflag=1;
            if(stopflag==1)
            {   EPwm1Regs.DBCTL.bit.OUT_MODE=0;
                EPwm2Regs.DBCTL.bit.OUT_MODE=0;
                EPwm3Regs.DBCTL.bit.OUT_MODE=0;
                EPwm1Regs.AQCSFRC.all=0x000A;
                EPwm2Regs.AQCSFRC.all=0x000A;
                EPwm3Regs.AQCSFRC.all=0x000A;
                rampspeed.SetpointValue=0;
                speedbldc.Ui =0;
                speedbldc.Fdb =0;
                speedbldc.Up =0;
                speedbldc.Out=0;
                currentq.Ui = 0;
                currentq.Up= 0;
                currentq.Fdb= 0;
                currentq.Out= 0;
                currentd.Ui = 0;
                currentd.Up=0;
                currentd.Fdb= 0;
                currentd.Out= 0;
                speed.Ui= 0;
```

```
                    speed.Up= 0;
                    speed.Fdb= 0;
                    speed.Out= 0;
                    ruoci.Ui= 0;
                    ruoci.Up= 0;
                    ruoci.Fdb= 0;
                    ruoci.Out= 0;
                }
                else if((abs(rdc.Speed)*rdc.BaseRpm)<50)
                {

                    EPwm1Regs.DBCTL.bit.OUT_MODE=0;
                    EPwm2Regs.DBCTL.bit.OUT_MODE=0;
                    EPwm3Regs.DBCTL.bit.OUT_MODE=0;
                    EPwm1Regs.AQCSFRC.all=0x000A;
                    EPwm2Regs.AQCSFRC.all=0x000A;
                    EPwm3Regs.AQCSFRC.all=0x000A;
                }
            }
        }
    }
}
```

References

1. X. Hu, T. Liu, X. Qi, M. Barth, Reinforcement learning for hybrid and plug-in hybrid electric vehicle energy management: recent advances and prospects. IEEE Ind. Electron. Mag. **13**(3), 16–25 (2019)
2. R. Zhang, E. Yao, Eco-driving at signalised intersections for electric vehicles. IET Intel. Transp. Syst. **9**(5), 488–497 (2015)
3. H. Cheng, Z. Wang, S. Yang, J. Huang, X. Ge, An integrated SRM powertrain topology for plug-in hybrid electric vehicles with multiple driving and onboard charging capabilities. IEEE Trans. Transp. Electrif. **6**(2), 578–591 (2020)
4. Y. Fang, T. Zhang, Sound quality investigation and improvement of an electric powertrain for electric vehicles. IEEE Trans. Industr. Electron. **65**(2), 1149–1157 (2018)
5. G.E.V. Outlook, *International Energy Agency* (France, Paris, 2019), p. 2019
6. G.E.V. Outlook, *International Energy Agency* (France, Paris, 2020), p. 2020
7. G.E.V. Outlook, *International Energy Agency* (France, Paris, 2018), p. 2018
8. Available at: http://www.cnelc.com/Article/50/160425/AD100394060_1.html
9. Available at: https://uk.reuters.com/article/us-tesla-autopilot/third-fatal-tesla-autopilot-crash-renews-questions-about-system-idUKKCN1SM1QE
10. F. Un-Noor, S. Padmanaban, L. Mihet-Popa, M.N. Mollah, E. Hossain, A comprehensive study of key electric vehicle (EV) components, technologies, challenges, impacts, and future direction of development. Energies **10**(8) (2017)
11. G. Liu, S. Chen, Z. Zeng et al., Risk factors for extremely serious road accidents: results from national road accident statistical annual report of China PLoS One **13**(8) (2018)
12. M. Nowakowska, K. Bęben, M. Pajęcki, A conceptual project for the organisation of road traffic crash and road traffic crash casualty data acquisition for research purposes, in *2019 International Conference on Information and Digital Technologies (IDT)*, Zilina, Slovakia, 2019, pp. 345–353
13. C. Visvikis, P. Morgan, P. Boulter et al., *Electric Vehicles: Review of Type-Approval Legislation and Potential Risks*, Transport Research Laboratory, Wokingham, the UK, Report CPR810, June 2010
14. C. Gong et al., Hybrid DC-Bus capacitor discharge strategy using internal windings and external bleeder for surface-mounted PMSM based EV powertrains in emergency. IEEE Trans. Industr. Electron. https://doi.org/10.1109/TIE.2020.2975479

C. Gong, *Crash Safety of High-Voltage Powertrain Based Electric Vehicles*, Springer Theses, https://doi.org/10.1007/978-3-030-88979-1

15. D. Naberezhnykh, W. Gillan, C. Visvikis, J. Cooper, M. Jones, *Implications of the Widespread Use of Electric Vehicles*, Transport Research Laboratory, Wokingham, the UK, TREL Insight Report INS010, 2011
16. M. Elkady, A. Elmarakbi, J. MacIntyre, M. Alhariri, Collision mitigation and vehicle transportation safety using integrated vehicle dynamics control systems. J. Traffic Transport. Eng. **4**(1), 41–60 (2017)
17. E. Krzystała, A. Mężyk, S. Kciuk, Minimisation of the explosion shock wave load onto the occupants Inside the vehicle during trinitrotoluene charge blast. Int. J. Inj. Contr. Saf. Promot. **23**(2), 170–177 (2014)
18. R. Chen, H.C. Gabler, Risk of thoracic injury from direct steering wheel impact in frontal crashes. J. Trauma Acute Care Surg. **76**(6), 1441–1446 (2014)
19. A.A. Żuchowski, Results of the crash tests of electric cars. J. KONES Powertrain Transp. **25**(1), 483–490 (2018)
20. R. Frampton, M. Page, P. Thomas, Factors related to fatal injury in frontal crashes involving European cars. Ann. Proc. Assoc. Adv. Autom. Med. **50**, 35–56 (2006)
21. M.A. Masrur et al., Military-based vehicle-to-grid and vehicle-to-vehicle microgrid-system architecture and implementation. IEEE Trans. Transport. Electrif. **4**(1), 157–171 (2018)
22. C. Zietz, G. Armbrecht, T. Schmid, M. Wollitzer, B. Geck, A general calibration procedure for measuring RF voltages and currents applied to the EMC analysis of automotive high-voltage power networks. IEEE Trans. Electromagn. Compat. **57**(5), 915–925 (2015)
23. Federal Motor Vehicle Safety Standards, No.305, Aug. 2015
24. T. Goldammer, T. Le, J. Miller, J. Wai, Active high voltage bus bleed down, U.S. Patent 20120161679 A1, Jun. 28, 2012
25. W. Yao, Q. Dai, Y. Liu, Simple chemical reduction route for the synthesis of Cu-modified Co_3O_4 nanosheet with enhanced performance as anode material in lithium-ion batteries. Micro Nano Lett. **13**(6), 784–787 (2018)
26. A. Rochliadi, B. Bundjali, I. M. Arcana, H. Dharmi, Polymer electrolyte membranes prepared by blending of poly (vinyl alcohol)-poly (ethylene oxide) for lithium battery application, in *Proceedings of the Joint International Conference on Electric Vehicular Technology and Industrial, Mechanical, Electrical and Chemical Engineering (ICEVT & IMECE)*, Surakarta, 2015, pp. 370–373
27. I.M. Arcana, B. Bundjali, A. Rochliadi, N.K. Hariyawati, Preparation of polymers electrolyte membranes from Styrofoam waste for lithium battery, in *2013 Joint International Conference on Rural Information & Communication Technology and Electric-Vehicle Technology (rICT & ICeV-T)*, Bandung, 2013, pp. 1–4
28. J. Zhang, L. Zhang, F. Sun, Z. Wang, An overview on thermal safety issues of lithium-ion batteries for electric vehicle application. IEEE Access **6**, 23848–23863 (2018)
29. Z.H. Che Daud, Z. Asus, S.A. Abu Bakar, N. Abu Husain, I.I. Mazali, D. Chrenko, Thermal characteristics of a lithium-ion battery used in a hybrid electric vehicle under various driving cycles. IET Electr. Syst. Transport. **10**(3), 243–248 (2020)
30. A. Kriston, I. Adanouj, V. Ruiz, A. Pfrang, Quantification and simulation of thermal decomposition reactions of Li-ion battery materials by simultaneous thermal analysis coupled with gas analysis. J. Power Sources **435**, 1–13 (2019)
31. A. Dutta, S. Debbarma, Contribution of electric vehicles for frequency regulation in presence of diverse power sources and transmission links, in *2018 IEEE International Conference on Industrial Technology (ICIT)*, Lyon, 2018, pp. 1177–1182
32. G.M. Madhu, C. Vyjayanthi, C.N. Modi, A novel framework for monitoring solar PV based electric vehicle community charging station and grid frequency regulation using blockchain, in *2019 10th International Conference on Computing, Communication and Networking Technologies (ICCCNT)*, Kanpur, India, 2019, pp. 1–7

33. P. Kunakron-ong, K. Ruangjirakit, P. Jongpradist, Design and analysis of electric bus structure in compliance with ECE safety regulations, in *2017 2nd IEEE International Conference on Intelligent Transportation Engineering (ICITE)*, Singapore, 2017, pp. 25–29
34. K. Xu, X. Wang, X. Meng, G. Liu, B. Wen, R. Zhou, Comparative research on standards and regulations of electric vehicle post crash safety requirement, in *2014 IEEE Conference and Expo Transportation Electrification Asia-Pacific (ITEC Asia-Pacific)*, Beijing, 2014, pp. 1–4
35. S. Fan et al., Current status and framework of China's inland passenger ship safety system, in *2016 International Conference on Logistics, Informatics and Service Sciences (LISS)*, Sydney, NSW, 2016, pp. 1–8
36. J. Gupta, B. Singh, A bidirectional home charging solution for an electric vehicle, in *2019 IEEE International Conference on Environment and Electrical Engineering and 2019 IEEE Industrial and Commercial Power Systems Europe (EEEIC / I&CPS Europe)*, Genova, Italy, 2019, pp. 1–6
37. S. Kim, S. Lee, S. Byun, Y. Cho, DC voltage regulation control of engine direct connection type IPMSG for special vehicle, in *2015 IEEE 15th International Conference on Environment and Electrical Engineering (EEEIC)*, Rome, 2015, pp. 1659–1664
38. United Nation Economic Commission for Europe Vehicle Regulation, No. 94 (ECE R94), Uniform provisions concerning the approval of vehicles with regard to the protection of the occupants in the event of a frontal collision, Rev. 2, Annex 11, Aug. 2013
39. Federal Motor Vehicle Safety Standards, No. 208, Oct. 1999
40. C. Visvikis, Safety considerations for electric vehicles and regulatory activities, in *EVS26 International Battery, Hybrid and Fuel Cell Electric Vehicle Symposium*, Los Angeles, California, 2012, pp. 1–13
41. R. Santos et al., Reliability-based design of automotive interior trimming for occupant safety: a case study, in *1st International Conference on Uncertainty in Structural Dynamics (USD 2007)*, Sheffield, the UK, 2007, pp. 1–14
42. S. Sudirja, A. Hapid, S. Kaleg, A.C. Budiman, Amin, The crumple zone quality enhancement of electric cars bumper fascia using a carbon fiber reinforced vinyl ester—microsphere composites, in *2019 International Conference on Sustainable Energy Engineering and Application (ICSEEA)*, Tangerang, Indonesia, 2019, pp. 1–5
43. G. Sudarshan, Structural optimization of automobile bumper using honeycomb structure. Int. J. Sci. Res. **7**(8), 969–978 (2018)
44. T. Niyazi, S. Fazil, S. Mahmut, Shape optimization of bumper beams under high-velocity impact loads. Eng. Struct. **95**, 49–60 (2015)
45. D. Kim, H. Kim, H. Kim, Design optimization and manufacture of hybrid glass/carbon fiber reinforced composite bumper beam for automobile vehicle. Compos. Struct. **131**, 742–752 (2015)
46. L. Zhao, J. Liu, P. Zhang, Lightweight design of automotive composite bumper system using modified particle swarm optimizer. Compos. Struct. **140**, 630–643 (2016)
47. F. Zeng, H. Xie, Q. Liu, F. Li, W. Tan, Design and optimization of a new composite bumper beam in high-speed frontal crashes. Struct. Multidiscip. Optim. **53**, 115–122 (2016)
48. L. Duan, H. Jiang, G. Geng, X. Zhang, Z. Li, Parametric modeling and multiobjective crashworthiness design optimization of a new front longitudinal beam. Struct. Multidiscip. Optim. **59**, 1789–1812 (2019)
49. H. Wang, H. Xie, Multi-objective optimization of crashworthiness of vehicle front longitudinal beam. Struct. Multidiscip. Optim. **61**, 2111–2123 (2020)
50. X. Chen, J. Yang, Optimization of Longitudinal Beam for Improvement of Crashworthiness in Frontal and Offset Impacts, in *2012 Third International Conference on Digital Manufacturing & Automation*, GuiLin, 2012, pp. 582–585.
51. C.N. Nguyen, T. Dirgantara, L. Gunawan, I.S. Putra, H.A. Ly, Analytical prediction of square crash box structure with holes due to impact loading, in *Regional Conference on Mechanical and Aerospace Technology*, Kuala Lumpur, 2013, pp. 1–10

52. J. Fu, M. Lei, Z. Lei, X. Luo, The CST design method of matching both European and American crash regulations, in *2011 Second International Conference on Mechanic Automation and Control Engineering*, Hohhot, 2011, pp. 778–781
53. L. Mi, X. Wei, X. Wan, M. Xia, A research on the impact energy-absorbing capacity of the aluminum alloy energy-absorbing box of the bumper. J. Chongqing Univ. Technol. (Nat. Sci.) **26**(6), 1–7 (2012). ((in Chinese))
54. D. Kim, J. Lim, B. Nam, H. Kim, H. Kim, Design and manufacture of automotive hybrid steel/carbon fiber composite B-Pillar component with high crashworthiness. Int. J. Precis. Eng. Manuf. Green Technol. (2020). https://doi.org/10.1007/s40684-020-00188-5
55. Z. Yang, Q. Peng, J. Yang, Lightweight design of B-pillar with TRB concept considering crashworthiness, in *2012 Third International Conference on Digital Manufacturing & Automation*, GuiLin, 2012, pp. 510–513
56. B. Gao, X. Chen, Z. Pan, A high-strength heterogeneous structural dual-phase steel. J. Mater. Sci. **54**(19), 1–13 (2019)
57. S. Zhang, *Research on Crashworthiness and Bionic Structural Design of Automobile Anti-Collision Beam*, Mater Thesis, School of Mechanical Engineering, Yanshan University, Qinhuangdao City, China, 2019
58. A. Sun, J. Liu, W. Liu, The laser-beam welding of ferritic stainless steel for the motor vehicle exhaust 409L, in *Proceedings of 2011 International Conference on Electronic & Mechanical Engineering and Information Technology*, Harbin, 2011, pp. 3931–3933
59. U.F. Shaikh, A. Das, A. Barai, I. Masters, Electro-thermo-mechanical behaviours of laser joints for electric vehicle battery interconnects, in *2019 Electric Vehicles International Conference (EV)*, Bucharest, Romania, 2019, pp. 1–6
60. A.E. Ikpe, E.K. Orhorhoro, A. Gobir, Design and reinforcement of a B-Pillar for occupants safety in conventional vehicle applications. Int. J. Math. Eng. Manage. Sci. **2**(1), 37–52 (2017)
61. A. Ghadianlou, S.B. Abdullah, Crashworthiness design of vehicle side door beams under low-speed pole side impacts. Thin-Walled Struct. **67**, 25–33 (2013)
62. M.A. Shaharuzaman, S.M. Sapuan, M.R. Mansor, M.Z.M. Yusoff, Passenger Car's side door impact beam: a review. J. Eng. Technol. **9**(1), 1–22 (2018)
63. F. Lu, D.-f. Wang, Structure optimization of door-beam base on strength of side doors, in *Proceedings 2011 International Conference on Transportation, Mechanical, and Electrical Engineering (TMEE)*, Changchun, 2011, pp. 766–769
64. R.W. Kent, S.V. Purtsezov, W.D. Pilkey, Limiting performance analysis of a seat belt system with slack. Int. J. Impact Eng **34**(8), 1382–1395 (2007)
65. J. Hu, M.P. Reed, J.D. Rupp, K. Fischer, P. Lange, A. Adler, Optimizing seat belt and airbag designs for rear seat occupant protection in frontal crashes. Stapp Car Crash J. **61**, 67–100 (2017)
66. A. Satija, P. Mishra, R. Gaurav, V. Singh, Load distribution optimization of seatbelt using validated finite element approach, in *SAE Technical Paper*, vol. 28, no. 2575, Nov. 2019
67. M. Oom, B.D. Jagar, Real-time optimal control of seat-belt systems. IFAC-Papers OnLine **50**(1), 10717–10722 (2017)
68. A.S. Kumar, S. Madhu, A research review on airbag in automobile safety system. Int. J. Appl. Eng. Res. **10**(33), 26815–26819 (2015)
69. H. Wang, H. Hong, J. Li, Q. Rui, Study on multi-objective optimization of airbag landing attenuation system for heavy airdrop. Defence Technol. **9**(4), 237–241 (2013)
70. H. Zhou, Z. Zhong, M. Hu, Design and occupant-protection performance analysis of a new tubular driver airbag. Engineering **4**(2), 291–297 (2018)
71. K. Cho, S.B. Choi, H. Lee, Design of an airbag deployment algorithm based on precrash information. IEEE Trans. Veh. Technol. **60**(4), 1438–1452 (2011)
72. X. Mei, H. Yang, X. Li, Y. Li, Y. Cheng, The effect of 5-Amino-1H-tetrazole on the combustion performance and ignition capability of boron/potassium nitrate igniter. J. Therm. Anal. Calorim. **120**, 1749–1754 (2015)

73. U. Nirmal, P.K. Ng, A comprehensive review on the development of car booster seats for children Current J. Appl. Sci. Technol. **38**(1) (2019)
74. S. Wen, Z. Zhao, W. Li, Q. Liang, S. Li, Z. Li, Composition for car interior trim and preparation method using the same. China Patent CN105936803B, Nov. 16, 2018
75. L. Xiang, Y. Xu, Automotive trim structure optimization based on finite element analysis considering crashworthiness. Shanghai Veh. **11**, 15–18 (2012). (in Chinese). 10.3969/j.issn.1007-4554.2012.11.04
76. F. Freschi, M. Mitolo, R. Tommasini, Electrical safety of plug-in electric vehicles: shielding the public from shock. IEEE Indus. Appl. Mag. **24**(3), 58–63 (2018)
77. F.J. Márquez-Fernández, S. Schuch, L. Lindgren, M. Alaküla, Electric safety challenges with a conductive electric road system—chassis potential modeling and measurement. World Electr. Veh. J. **10**(30), 1–10 (2019)
78. A. Christe, E. Coulinge, D. Dujic, Insulation coordination for a modular multilevel converter prototype, in *2016 18th European Conference on Power Electronics and Applications (EPE'16 ECCE Europe)*, Karlsruhe, 2016, pp. 1–9
79. Y.H. Kim et al., Dielectric characteristics of solid insulation materials with respect to surface roughness. IEEE Trans. Appl. Superconduct. **25**(3), 1–4 (2015). Art no. 7700404
80. H. Kim, J. Cho, W. Kim, Y.S. Choi, S. Kim, The basic dielectric characteristics of insulating materials for HTS DC cable system. IEEE Trans. Appl. Superconduct. **26**(3), 1–4 (2016). Art no. 7701104
81. G. Jegatheesh, R. Muniraj, Fabrication and analysis on critical parameters of nanosolid dielectric material for enhancing the insulation strength, in *2015 International Conference on Innovations in Information, Embedded and Communication Systems (ICIIECS)*, Coimbatore, 2015, pp. 1–5
82. H. Dai, X. Zhang, X. Wei, Z. Sun, High voltage safety analysis and protection design in EV development. Appl. Commun. **1**, 53–59 (2013). ((in Chinese))
83. X. Zhang, W. Ren, S. Wang, Experimental investigation of contact welding characteristics of electromechanical relays. IEEE Trans. Indust. Electron. https://doi.org/10.1109/TIE.2020.3009565
84. G. Uhl, Dr.-Ing, Electronic circuit breaker for battery, European Patent EP1600337A1, Sep. 11, 2008
85. N. Paul, T. Gopal, Pyrotechnic high voltage battery disconnect, United States Patent US9221343B2, Dec. 29, 2015
86. P. Sun, R. Bisschop, H. Niu, X. Huang, A review of battery fires in electric vehicles. Fire Technol. **56**, 1–33 (2020)
87. C. Cope, *Battery Disconnect Devices Related to Vehicle Safety*. 2008 ASME International Mechanical Engineering Congress and Exposition, Boston, Massachusetts, USA, 2008, pp. 83–92
88. S. Liu, J. Wu, W. Yang et al., *A Protective Disconnect System for Electric Vehicles in Crash*, Chines Patent, CN104494544B, Jan. 18, 2017
89. S. Ashida, K. Yamada, M. Nakamura, T. Shimana, T. Soma, *Electric Vehicle, and Control Apparatus and Control Method for Electric Vehicle*, U.S. Patent 8631894 B2, Jan. 21, 2014
90. J.W. Meyer, D.P. Tasky, S.M. Nayeem Hasan, B.M. Ludwig, *Passive High-Voltage DC Bus Discharge Circuit for a Vehicle*, U.S. Patent 9018865 B2, Apr. 28, 2015
91. D.J. Kaplan, *High-Voltage Bus Discharge with Logarithmic Self-Protection*, U.S. Patent 20090268354 A1, Oct. 29, 2009
92. H. Peng, J. Wang, W. Shen, D. Shi, Y. Huang, Controllable regenerative braking process for hybrid battery–ultracapacitor electric drive systems. IET Power Electron. **11**(15), 2507–2514 (2018)
93. E. Ganev, L.A. Fizer, *DC Bus Short Circuit Compliant Power Generation Systems Using Induction Machine*, U.S. Patent 459889 B2, Oct. 29, 2009

94. Z. Wu, X. Su, Y. Zhu, M. Xiao, DC link capacitor active discharge by IGBT weak short circuit. SAE Int. J. Adv. Current Pract. Mob. **1**(3), 1177–1187 (2019)
95. S. Hiti, D. Tang, B.A. Welchko, M. Perisic, C.C. Stancu, *Redundant DC Bus Discharge for an Electric motor System*, U.S. Patent 7944160 B2, Sep. 17, 2009
96. C. Gong, Y. Hu, G. Chen, H. Wen, Z. Wang, K. Ni, A DC-bus capacitor discharge strategy for PMSM drive system with large inertia and small system safe current in EVs. IEEE Trans. Industr. Inf. **15**(8), 4709–4718 (2019)
97. Z. Ke, J. Zhang, M.W. Degner, DC bus capacitor discharge of permanent-magnet synchronous machine drive systems for hybrid electric vehicles. IEEE Trans. Industr. Appl. **53**(2), 1399–1405 (2017)
98. J. Istiyanto, D.A. Sumarsono, M.S. Utomo, G. Kiswanto, A.S. Baskoro, S. Supriadi, Development of platform structure as protection to lithium batteries in electric vehicle during crash impact, in *2013 Joint International Conference on Rural Information & Communication Technology and Electric-Vehicle Technology (rICT & ICeV-T)*, Bandung, 2013, pp. 1–8
99. R. Justen, R. Schöneburg, Crash safety of hybrid and battery electric vehicles, in *22nd International Technical Conference on the Enhanced Safety of Vehicles (ESV)*, Washington DC, 2011
100. Y. Lian, D. Zeng, S. Ye et al., High-voltage safety improvement design for electric vehicle in rear impact. Autom. Innov. **1**, 211–225 (2018)
101. Z. Wang, S. Shi, P. Liu, Research progress on collision safety of electric vehicles, in *2011 Third International Conference on Measuring Technology and Mechatronics Automation*, vol. 3, 2011, pp. 153–156
102. N. Bao, R. Zhao, Design optimization of battery holder for electric vehicle, in *2018 6th International Conference on Mechanical, Automotive and Materials Engineering (CMAME)*, Hong Kong, 2018, pp. 79–84
103. S. Li, F. Chen, A. Garg, X. Peng, N. Bao, J. Zhang, Design optimization of battery pack enclosure for electric vehicle. Struct. Multidiscip. Optim. **58**, 331–347 (2018)
104. C. Suh, K. Hor, S. Nahm, M. Suh, Fatigue and material characteristics of a hot-formed AZ31 magnesium alloy. Modern Phys. Lett. B **29**(6, 7), (2015)
105. Y. Zhang, Y. Cao, X. Hong et al., Study on lightweight for electric car battery case. Modern Manuf. Eng. **1**, 38–41 (2014). ((in Chinese))
106. K. Saeki, *Battery Protection Structure for Automobile*, U.S. Patent, US8708402B2, Apr. 29, 2014
107. Y. Xia, T. Wierzbicki, E. Sahraei, X. Zhang, Damage of cells and battery packs due to ground impact. J. Power Sour. **267**(1), 78–97 (2014)
108. X. Zhang, T. Wierzbicki, Characterization of plasticity and fracture of shell casing of lithium-ion cylindrical battery. J. Power Sour. **280**(15), 47–56 (2015)
109. H. Fadillah, S.P. Santosa, A. Jusuf, T. Dirgantara, Li-ion NCA battery safety assessment for electric vehicle applications, in *2018 5th International Conference on Electric Vehicular Technology (ICEVT)*, Surakarta, Indonesia, 2018, pp. 172–178
110. Y. Xia, T. Wierzbicki, E. Sahraei, X. Zhang, Damage of cells and battery packs due to ground impact. J. Power Sour. **267**, 78–97 (2014)
111. J. Zhu, X. Zhang, T. Wierzbicki, Y. Xia, G. Chen, Structural designs for electric vehicle battery pack against ground impact, in *Presented at the WCX World Congress Experience*, 2018
112. P.N. Halimah, S.P. Santosa, A. Jusuf, T. Dirgantara, The concept of sandwich panel structures for battery protections in electric vehicles subjected to ground impact, in *2018 5th International Conference on Electric Vehicular Technology (ICEVT)*, 2018, pp. 142–146
113. T. Nirmala, A. Jusuf, S. Puji Santosa, P.L. Sambegoro, Design study of battery system protection structure based on hybrid material fiber metal laminate (FML), in *2019 6th International Conference on Electric Vehicular Technology (ICEVT)*, Bali, Indonesia, 2019, pp. 163–169

114. S. Basak, S.K. Panda, Y.N. Zhou, Formability assessment of prestrained automotive grade steel sheets using stress based and polar effective plastic strain-forming limit diagram. J. Eng. Mater. Technol. **137**(4), 041006–041017 (2015)
115. W. Yang, A.A. Pelegri, Numerical evaluation of stiffness and energy absorption of a hybrid unidirectional/random glass fiber composite. J. Eng. Mater. Technol. **133**(4), 041018–041025 (2011)
116. X. Ding et al., A novel active equalization topology for series-connected lithium-ion battery packs. IEEE Trans. Indus. Appl. https://doi.org/10.1109/TIA.2020.3015820
117. G.J. Su, L. Tang, A segmented traction drive system with a small dc bus capacitor, in *2012 IEEE Energy Conversion Congress and Exposition*, Raleigh, NC, 2012, pp. 2847–2853
118. Z. Mynar, L. Vesely, P. Vaclavek, PMSM model predictive control with field-weakening implementation. IEEE Trans. Industr. Electron. **63**(8), 5156–5166 (2016)
119. T. Tarczewski, L.M. Grzesiak, Constrained state feedback speed control of PMSM based on model predictive approach. IEEE Trans. Industr. Electron. **63**(6), 3867–3875 (2016)
120. J. Linares-Flores, C. García-Rodríguez, H. Sira-Ramírez, O.D. Ramírez-Cárdenas, Robust backstepping tracking controller for low-speed PMSM positioning system: design, analysis, and implementation. IEEE Trans. Industr. Inf. **11**(5), 1130–1141 (2015)
121. G. Feng, C. Lai, N.C. Kar, Particle-filter-based magnet flux linkage estimation for PMSM magnet condition monitoring using harmonics in machine speed. IEEE Trans. Industr. Inf. **13**(3), 1280–1290 (2017)
122. J. Liu, C. Gong, Z. Han, H. Yu, IPMSM model predictive control in flux-weakening operation using an improved algorithm. IEEE Trans. Ind. Electron. **65**(12), 9378–9387 (2018)
123. A. Rabiei, T. Thiringer, M. Alatalo, E.A. Grunditz, Improved maximum-torque-per-ampere algorithm accounting for core saturation, cross-coupling effect, and temperature for a PMSM intended for vehicular applications. IEEE Trans. Transp. Electrif. **2**(2), 150–159 (2016)
124. H. Ge, B. Bilgin, A. Emadi, Global loss minimization control of PMSM considering cross-coupling and saturation, in *2015 IEEE Energy Conversion Congress and Exposition*, Montreal, QC, 2015, pp. 6139–6144
125. L. Sun, X. Wang, Z. Deng, Q. Ding, Research on stable regions of double-closed loop system for PMSM with low-pass filters, in *2014 17th International Conference on Electrical Machines and Systems*, Hangzhou, 2014, pp. 1150–1156
126. C.-Z. Liaw, W.L. Soong, B.A. Welchko, N. Ertugrul, Uncontrolled generation in interior permanent-magnet machines. IEEE Trans. Indus. Appl. **41**(4), 945–954 (2005)
127. S. Xiaoxia, W. Yichun, L. Meng, F. Wang, L. Xiang, Research on braking resistor of hybrid electric armored vehicle, in *2010 International Conference on Information, Networking and Automation (ICINA)*, Kunming, 2010, pp. V2-206–V2-210
128. C. Gong, Y. Hu, C. Gan, G. Chen, M. Alkahtani, Modeling, analysis, and attenuation of uncontrolled generation for IPMSM-based electric vehicles in emergency. IEEE Trans. Industr. Electron. **67**(6), 4453–4462 (2020)
129. Z. Zhang, X. Zou, Z. Wu, Y. Zou, Analysis on harmonic current of three-phase bridge uncontrolled rectifier, in *2010 Asia-Pacific Power and Energy Engineering Conference*, Chengdu, 2010, pp. 1–4
130. P.L. Xu, Z.Q. Zhu, Novel carrier signal injection method using zero-sequence voltage for sensorless control of PMSM drives. IEEE Trans. Industr. Electron. **63**(4), 2053–2061 (2016)
131. S.K. Kommuri, S.B. Lee, K.C. Veluvolu, Robust sensors-fault-tolerance with sliding mode estimation and control for PMSM drives. IEEE/ASME Trans. Mechatron. **23**(1), 17–28 (2018)
132. G. Zhang, G. Wang, D. Xu, R. Ni, C. Jia, Multiple-AVF cross-feedback-network-based position error harmonic fluctuation elimination for sensorless IPMSM drives. IEEE Trans. Industr. Electron. **63**(2), 821–831 (2016)

133. M. Bouheraoua, J. Wang, K. Atallah, Rotor position estimation of a pseudo direct-drive PM machine using extended Kalman filter. IEEE Trans. Indus. Appl. **53**(2), 1088–1095 (2017)
134. J. Zhao, M. Netto, L. Mili, A robust iterated extended Kalman filter for power system dynamic state estimation. IEEE Trans. Power Syst. **32**(4), 3205–3216 (2017)
135. S.K. Kommuri, S.B. Lee, K.C. Veluvolu, Robust sensors-fault-tolerance with sliding mode estimation and control for PMSM drives. IEEE/ASME Trans. Mechatron. **23**(1), 17–28 (2018)
136. X. Song, J. Fang, B. Han, S. Zheng, Adaptive compensation method for high-speed surface PMSM sensorless drives of EMF-based position estimation error. IEEE Trans. Power Electron. **31**(2), 1438–1449 (2016)
137. M. Golestani, P. Ahmadi, A. Fakharian, Fractional order sliding mode guidance law: improving performance and robustness, in *2016 4th International Conference on Control, Instrumentation, and Automation (ICCIA)*, Qazvin, 2016, pp. 469–474
138. M. Van, S.S. Ge, H. Ren, Robust fault-tolerant control for a class of second-order nonlinear systems using an adaptive third-order sliding mode control. IEEE Trans. Syst. Man Cybern. Syst. **47**(2), 221–228 (2017)
139. C. Gong, Y. Hu, J. Gao, Y. Wang, L. Yan, An improved delay-suppressed sliding-mode observer for sensorless vector-controlled PMSM. IEEE Trans. Industr. Electron. **67**(7), 5913–5923 (2020)
140. J. Lu, X. Zhang, Y. Hu, J. Liu, C. Gan, Z. Wang, Independent phase current reconstruction strategy for IPMSM sensorless control without using null switching states. IEEE Trans. Industr. Electron. **65**(6), 4492–4502 (2018)
141. G. Wang, L. Yang, G. Zhang, X. Zhang, D. Xu, Comparative investigation of pseudorandom high-frequency signal injection schemes for sensorless IPMSM drives. IEEE Trans. Power Electron. **32**(3), 2123–2132 (2017)
142. S. Medjmadj, D. Diallo, M. Mostefai, C. Delpha, A. Arias, PMSM drive position estimation: contribution to the high-frequency injection voltage selection issue. IEEE Trans. Energy Convers. **30**(1), 349–358 (2015)
143. R. Ni, D. Xu, F. Blaabjerg, K. Lu, G. Wang, G. Zhang, Square-wave voltage injection algorithm for PMSM position sensorless control with high robustness to voltage errors. IEEE Trans. Power Electron. **32**(7), 5425–5437 (2017)
144. D. Kim, Y. Kwon, S. Sul, J. Kim, R. Yu, Suppression of injection voltage disturbance for high-frequency square-wave injection sensorless drive with regulation of induced high-frequency current ripple. IEEE Trans. Indus. Appl. **52**(1), 302–312 (2016)
145. Y. Yoon, S. Sul, Sensorless control for induction machines based on square-wave voltage injection. IEEE Trans. Power Electron. **29**(7), 3637–3645 (2014)
146. A. Chalanga, S. Kamal, L.M. Fridman, B. Bandyopadhyay, J.A. Moreno, Implementation of super-twisting control: super-twisting and higher order sliding-mode observer-based approaches. IEEE Trans. Industr. Electron. **63**(6), 3677–3685 (2016)
147. D. Liang, J. Li, R. Qu, W. Kong, Adaptive second-order sliding-mode observer for PMSM sensorless control considering VSI nonlinearity. IEEE Trans. Power Electron. **33**(10), 8994–9004 (2018)
148. Y. Kwon, S. Sul, Reduction of injection voltage in signal injection sensorless drives using a capacitor-integrated inverter. IEEE Trans. Power Electron. **32**(8), 6261–6274 (2017)
149. A. Lidozzi, L. Solero, F. Crescimbini, A. Di Napoli, Direct tuning strategy for speed controlled PMSM drives, in *2010 IEEE International Symposium on Industrial Electronics*, Bari, 2010, pp. 1265–1270
150. L. Wang, H. Zhu, Z. Yang, Tuning method for PI controllers of PMSM driving system. Trans. China Electrotech. Soc. **29**(05), 104–117 (2014)
151. R.C. Dorf, R.H. Bishop, *Modern Control Systems*, 12th ed. (Prentice Hall, New York, 2010)